Diamond Jubilee Historical/Review Volume

AIChE Symposium Series

Larry Resen, editor

235 *Volume 80, 1983*

AMERICAN INSTITUTE OF CHEMICAL ENGINEERS

ERRATUM

DIAMOND JUBILEE HISTORICAL/REVIEW VOLUME

AIChE SYMPOSIUM SERIES ISSUE NO. 235

should read **Volume 80** not Volume 79

Diamond Jubilee
Historical/Review
Volume

Larry Resen, editor

N.R. Amundson
M. Benedict
J.A. Buckham
P.S. Buckley
S.W. Churchill
J.S. Dranoff
J.F. Fair

C.D. Holland
H.M. Hulburt
E.A. Mason
D.F. Othmer
S.I. Proctor
R.W.H. Sargent
R. Shinnar

AIChE Symposium Series

Number 235   1983   Volume 30

Published by

American Institute of Chemical Engineers

45 East 47 Street   New York, New York 10017

Copyright 1983

American Institute of Chemical Engineers
345 East 47 Street, New York, N.Y. 10017

AIChE shall not be responsible for statements or opinions advanced in papers or printed in its publications.

Library of Congress Cataloging in Publication Data
Main entry under title:

Diamond jubilee historical/review volume.

(AIChE symposium series ; no. 235, v. 80 79)
Papers presented at the AIChE diamond jubilee meeting in Washingtn, D.C.,
in the fall of 1983.
  1. Chemical engineering—Congresses. I. Resen, Larry. II. American
Institute of Chemical Engineers. Meeting (75th : 1983 : Washington, D.C.)
III. Series : AIChE Symposium series ; no. 235.
TP5.D53 1983   660.2   84-11155
ISBN 0-8169-0323-9   87-11141

Printed in the United States of America by
Twin Production & Design

FOREWORD

At the Diamond Jubilee Meeting of the AIChE in Washington, DC, Fall, 1983, a stellar program and array of papers was presented. Many of them have been published in various Institute periodicals. Among those presented were those included in this volume which were selected for their overview of various segments of technology. The main regret an editor has in making selections is how many excellent treatises must be omitted due to budgetary reasons. It is unfortunate it was not possible to publish a Proceedings of the entire program.

One interesting group of papers in this volume are those presented on aspects of the nuclear industry by former winners of the Robert E. Wilson Award in Nuclear Engineering. A notable one that is not in this volume may be found in the June, 1984, issue of *Chemical Engineering Progress*. The author: the Hon. W. Kenneth Davis, Past President of AIChE.

Finally, since this is the only direct publication relating to the Diamond Jubilee Meeting beyond the January, February and March, 1984, issues of *CEP*, included herein is the list of Eminent Chemical Engineers, selected by their peers for recognition at the meeting.

Larry Resen, *editor*
American Institute of Chemical Engineers
New York, New York

AMERICAN INSTITUTE OF
CHEMICAL ENGINEERS

1908 / 1983

Eminent
Chemical
Engineers

1983 Annual Meeting & Diamond Jubilee
The Washington Hilton, Washington, D.C.
October 30–November 4, 1983

W. Kenneth Davis
Thomas B. Drew
Harry G. Drickamer
James R. Fair
George E. Holbrook
Hoyt C. Hottel
Olaf A. Hougen
Arthur E. Humphrey
Donald L. Katz
Chalmer G. Kirkbride
Ralph Landau
W. Robert Marshall
Jerry McAfee
John J. McKetta, Jr.
Arthur B. Metzner
James Y. Oldshue
Max S. Peters
Robert L. Pigford
J. Henry Rushton
Klaus D. Timmerhaus
James Wei
James W. Westwater
Charles R. Wilke

Neal R. Amundson
Thomas Baron
Manson Benedict
R. Byron Bird
Theodore A. Burtis
Stuart W. Churchill
Donald A. Dahlstrom

Thirty Eminent Chemical Engineers Were Singled Out
by a special committee to honor them for their professional achievements.

CONTENTS

HISTORICAL DEVELOPMENT OF DISTILLATION EQUIPMENT

James R. Fair ■ Department of Chemical Engineering, The University of Texas, Austin, TX 78712

The ancient unit operation of distillation was of fundamental importance to the alchemists and was widely practiced before the time of Christ. In its more modern, multi-stage form, distillation dates back to the early part of the nineteenth century. In this paper the evolution of column-type distillation equipment is traced in detail from about 1830 to the present time. Developments have continued without pause and, because of the strategic importance of distillation in separations technology, may be expected to continue well into the future.

Distillation is indeed an ancient unit operation. It was of fundamental importance to the alchemists, and was widely practiced before the time of Christ. Separations were carried out in crude vaporization/condensation equipment, often for the purpose of increasing the alcoholic content of wine. Illustrations of early stills have been preserved and may be found in the treatises of Underwood (73) and Forbes (25). Some early examples are shown in Figures 1 and 2, the latter indicating a provision for addition of steam in the distillation of gums and resins. Some of the later developments of these takeover devices included provisions for staging, and re-distillation was early discovered as a means for obtaining more concentrated products. Ultimately there developed the concept of vertical, columnar devices in which descending liquid and ascending vapor could be brought into intimate contact, and which satisfied the production needs for continuous operation.

The first vertical, columnar configuration of continuous distillation equipment appears to have been developed by Cellier-Blumenthal in 1813 and 1818 (10). His 1818 version is shown in Figure 3. Shortly afterward Perrier (53) introduced an early version of the bubble-cap dispersing device (Figure 4), and Cellier-Blumenthal in 1834 showed a tray layout of bubble caps that looks very much like mid-twentieth century fractionator design practice (Figure 5). The first sieve tray column was developed by Coffey in 1830 (15) and Figure 6 shows his arrangement that includes feed

preheating by exchange with the bottoms stream and with condensing distillate. The column contained an internal condenser for providing reflux and counterflow contacting in a rectifying zone. The Coffey column even contained "several light valves opening upwards so that vapour can pass upwards through them" (73). Another feature of Coffey's trays was that instead of downpipes for liquid flow a portion of the tray metal was bent downward to form the equivalent of the more modern segmental downcomer baffle. It seems safe to conclude that the modern version of distilling columns had its origin with the devices of Cellier-Blumenthal and Coffey. Another Coffey still, developed some years later, is shown in Figure 7.

As is well known, distillation columns are identified according to their type of internal contacting devices. Trays provide staged-type contacts, and the liquid is usually in crossflow to the ascending vapors. Packings operate in a counterflow mode and there is continuous contacting of vapor and liquid. According to Underwood (73), packings were used as early as 1820 by a technologist named Clement who, drawing on the inventions of Cellier-Blumenthal, used glass balls of about 25 mm diameter in an alcohol still. Underwood also mentioned an 1847 patent by Phillips which described a still column filled with coke or pumice stone to effect interaction between vapor and liquid. In 1881 Hempel introduced packed columns at the laboratory scale, and used 4 mm glass beads. Before 1890 a person

1

named Ilges used 25 to 50 mm porcelain balls in plant scale equipment (46). The use of spherical packing elements made possible dumped beds with reasonable liquid distribution characteristics, but it was discovered quite early, especially as various odd-shaped elements were proposed and used, that there was great danger of liquid and vapor channeling. This tended to delay the effective use of packings in larger diameter distillation columns.*

At the turn of the last century the art of distillation was beginning to change to a science. In the 1890s Sorel (67) and Hausbrand (31) had published methods for determining required stages of contact for a given separation. The bubble-cap and sieve tray columns had become established as the chemical industry developed, and it was common to employ distillation columns with diameters up to a meter or more.

THE PERIOD 1900-1925

The first quarter of the present century found considerable expansion in the application of tray columns, particularly as the petroleum industry developed. Practice appears to have favored bubble-cap trays, but notable exceptions, when severe corrosion was indicated, included absorption towers filled with packing elements of chemical stoneware. It was during the latter part of this period that special packing elements such as Raschig rings, Berl saddles and Lessing rings were introduced (9). Some examples of various packing elements are shown in Figure 8.

Apart from the equipment aspect, notable advances were made in the evaluation of height requirements for distillation columns. The two-film theory of mass transfer was proposed in 1923 by Whitman (74), the concept of stage efficiency by Lewis in 1922 (43), the equivalence of the height of a packed column section to one theoretical tray (the "HETP") by Peters in 1922 (54), and the graphical determination of theoretical stages by McCabe and Thiele in 1925 (48).

THE PERIOD 1925-1950

In a 1933 state-of-the-art review on fractional distillation, Leslie and Coates (38) noted that

---

*In 1934 Carey pointed out that "packed columns have not found widespread use . . . . in sizes larger than about 20 inches [51 cm] in diameter owing to the difficulty in maintaining good reflux distribution." (9)

Interest in the subject has greatly increased during the last ten years as a result of widening use of fractionation and, particularly, the awakening technical consciousness in the petroleum and natural gas industries.

They added, however, that

Data on plate efficiency are too few. . . Little of a specific nature has been written on the subject of column capacity. Fairly dependable data are in possession of users and suppliers of column equipment, but few of these have been published. . . although [columns of the cap type] are by far the most important, the packed column is not without its field of usefulness. . . However, experience has shown that packed columns are not efficient if built in large diameter because of channeling of the fluids flowing. . . The usefulness of the packed column is limited, and it is hardly to be expected that it will be greatly extended.

During this period, bubble caps predominated as devices for dispersing gas into the tray liquid, and the exigencies of wartime production in large equipment fostered studies of the hydraulic characteristics of large diameter trays. There were a number of ideas put forward on the optimum design of caps and, especially, their slot configurations; some variations are shown in Figure 9. Approaches for dealing with the design and operation of large columns during World War II were detailed by Harrington et al.(30) in an article titled "No Peace for Fractionators", and in an effort to share experiences across the industry the Technical Advisory Committee of the Petroleum Administrator for War was set up. This committee published the first comprehensive design method for bubble-cap tray columns.(14) One of the co-authors of this work, J. A. Davies, published a more directly useful and obtainable design procedure in 1947 (19); this work became essentially a standard method in many design offices. Another useful piece of work represented the first published results of controlled experiments on the hydraulic characteristics of bubble-cap trays (29). In addition, the thesis work of Dauphine' (18) on pressure drop for flow of gases through bubble caps led to improved understanding of the effect of cap geometry on the liquid capacity of downcomers.

It is noteworthy that during this period sieve trays were essentially shunned by process engineers. Apparently some of the earlier designs of these trays called for open areas for gas flow that were much larger than they should have been, with the result of heavy weeping of liquid and poor flexibility for varying loadings. There were advocates of sieve trays in the late 1940's, but the known reliability of bubble-cap trays outweighed heavily the indicated capital cost advantages of the sieve trays. At the end of the period Hutchinson et al. (34) reported on sieve tray hydraulics studies and fashioned several of the concepts of froth aeration that led to rational design procedures for such trays. The Hutchinson et al. paper was never published in a journal, but it has been distributed widely by various means.

Several notable publications supported this growth of understanding of the behavior of bubble-cap trays. The book, "Elements of Fractional Distillation" by C. S. Robinson(58), first published in 1922, was revised in 1930, 1939 and 1950, the last time by E. R. Gilliland (59); the last edition remains as something of a classic in the distillation literature. Murphree had earlier defined overall tray efficiency (51); Lewis in 1936 showed how this efficiency related to that at any point on the tray (45). Colburn in 1936 derived equations to show the deleterious effect of liquid entrainment on the efficiency of trays (16). Souders and Brown (68) developed a rational basis for determining column diameter and tray spacing as a function of allowable vapor velocity. And in the area of equilibrium stage calculations, the famous papers by Lewis and Matheson (44) and by Thiele and Geddes (70) led the way to current advanced methods for computing rigorously the equilibrium stage requirements for separating multicomponent mixtures. Another important paper, not so recognized at the time, was published by Gerster and co-workers at the University of Delaware (28). These investigators showed how transfer unit concepts could be applied to the analysis and mass transfer prediction of bubble-cap trays. This Delaware work continued and later formed the basis for an industry sponsored distillation tray efficiency research program.

In summary, at the end of this quarter century it had become possible to execute careful and reliable designs of bubble-cap trays, including those of very large size containing two or more liquid flow passes. The stage was set for optimization studies of tray layouts. As things turned out, the stage was set also for the introduction of a number of contacting devices that would compete successfully with the traditional bubble-cap tray.

THE PERIOD 1950-1975

There were so many important developments during this next quarter century that it is appropriate to subdivide the discussion into parts relating to classes of contacting devices. The chronology of developments will thus be handled in a parallel fashion.

Bubble-Cap Trays

As noted, the 1940's saw many improvements in the understanding of bubble-cap tray behavior. These related primarily to hydraulics, but they provided a solid base for study of the mass transfer efficiency of cap trays as well as other crossflow type devices. The Delaware paper of 1948, mentioned above, led to the formation of an industry-sponsored research program on tray efficiency, coordinated by the AIChE Research Committee. Experimental studies were conducted at the Universities of Delaware and Michigan and at North Carolina State University. The program began in 1952 and ran for five years, being funded by some forty industrial organizations. Final reports were issued by the three universities, and the recommended methods for analysis and design are contained in an available design manual (1). This manual, together with a comprehensive treatment by Bolles (5)stand today as the "last word" on bubble-cap tray design.

Changes in the mechanical design of cap trays became evident. Whereas in the pre-1940 era most trays and caps were of cast iron, the newer trays and caps were formed from sheet metal. Pipe downspouts gave way to downcomers formed from simple segmental baffles. Massive I-beam supports for the heavy cap load (Figure 10) were often replaced by integral truss-type members developed by the Glitsch company. Work at the University of Delaware, extended later by Fractionation Research, Inc., showed that the design of slots for the caps was not critical, and the "teacup cap", having no slots at all, emerged.

Mention of Fractionation Research, Inc. (FRI) is a reminder that this industry-sponsored private research company was formed in 1952 for the purpose of testing fractionating devices as well as entire distillation systems. The initial sponsors were primarily petroleum companies and engineering design firms. After its experimental program got underway in about 1954, FRI placed emphasis on bubble-cap tray testing. More will be said about FRI which, in a very real sense, has had a major influence on the selection /

specification of distillation column contacting devices in the last quarter century.

During this period variants of the standard bell-type tray were introduced. These included the Uniflux tray (8), shown in Figure 11. Muller and Othmer (50) provided useful Uniflux tray operating data, obtained in the FRI test facilities. Another FRI-tested tray was that of Thormann (57), a tunnel-cap affair designed to encourage plug flow of liquid across the tray (and thus enhance efficiency). These and similar devices had relatively short lives because they did not materially improve on the performance of the better-known bell cap trays.

By the mid-1960's very few new designs called for the use of bubble caps. Complete inroads had been made by valve trays, sieve trays and, in some cases, special packings. By 1975 the bubble-cap tray was a relic of the past, so far as new designs were concerned. However, there are thousands of bubble-cap columns still in operation, so the technology of their performance is still a matter of interest.

Counterflow Trays

In an effort to use more fully the total tower area for flow of the phases, a "downcomerless" perforated tray was developed on the basis that an individual perforation would pass both vapor and liquid, but not both at the same time. The early still of Cellier-Blumenthal utilized this principle. At the time of introduction in the early 1950's it was probably not recognized that each perforation would perform the dual service, but this turned out to be a unique advantage of the counterflow tray since it permitted the perforations to remain clean when dirty liquids were being handled. In 1952 Shell Development introduce the turbogrid tray in which the openings were in the form of narrow rectangular slots (63). This tray was later found to require de-rating at larger diameters (76). One form of a turbogrid tray is shown in Figure 12. An equivalent tray, having round openings and called the dualflow tray (Figure 13) was tested extensively by FRI. Stone and Webster Engineering Corporation introduced the ripple tray (33) which was identical to the dualflow tray except that the tray metal was corrugated. The self-cleaning features of this tray were also recognized, with excellent results reported in the processing of fermentate for alcohol recovery; hole sizes of 10 mm were large enough to pass the solids contained in the feed to the beer still. Another counterflow tray, introduced in the United States in 1953 (55) but developed earlier in West Germany, was the Kittle tray. This device utilized relatively inexpensive expanded metal as tray material.

Older devices that could be considered counterflow in their contacting action were the baffle tray columns, utilizing simple segmental baffles (sometimes partially perforated and called "shower decks") or disc-and-doughnut baffles. These columns were found to perform well in dirty service, except for their low mass transfer efficiencies. A variant on the baffle column, designed for improved fluid contacting, was the Schneible tower, patented in 1952 (62).

Counterflow tray devices have survived in various forms to the present day, but are limited mostly to services for which they excel: handling of dirty liquids that would foul or plug conventional crossflow trays and packings.

Sieve Trays

The 1949 paper by Hutchinson and co-workers, mentioned earlier, appeared to bolster the confidence of process engineers in the usefulness of the crossflow sieve tray. Another pioneering paper, by Mayfield and co-workers (47) provided large-scale air/water simulator data for a variety of sieve tray designs as well as performance data from a sieve tray installation at Celanese Corporation. Initially, hole sizes were limited to the 3 to 6 mm range, but later work by FRI showed that good results could be obtained with hole sizes as large as 12 mm. By the mid-1950's a number of sieve tray columns were in service throughout the United States. Continuing work in West Germany was summarized by Kirschbaum (37) and this furnished additional input for designers. Articles by Hughmark and O"Connell (32) and Fair (23) provided complete design procedures. The sieve tray continued to gain favor because it gave high efficiencies and it was the least expensive of the crossflow devices to fabricate. A typical sieve tray layout is shown in Figure 14.

One problem with the specification of sieve trays was the lack of performance data for columns of commercial size. Papers by Jones and Pyle (35), Rush and Stirba (60), Kastanek and Standart (36), Zuiderweg et al. (75) and Billet et al. (4) contained useful efficiency data, as did the book by Kirschbaum (37). But it was not until 1979 that large-scale data on sieve trays were released by FRI (61). A recent paper by Chan and Fair (11) includes a review of published data on large scale sieve trays as well as a generalized method for sieve tray efficiency prediction.

Some variations on the conventional

crossflow sieve tray have been introduced. The Esso jet tray (26) utilized tabs to direct vapor flow in the direction of liquid flow. The Linde subsidiary of Union Carbide developed a directionally-slotted sieve tray (27) and a multiple-downcomer tray (20) with considerable success. Among other things, these trays were designed to eliminate stagnant, unproductive zones on trays; such zones could result from poorly distributed crossflowing liquid.

In many respects the sieve tray has become the standard of comparison for tray-type columns. It is simple, relatively easy to model, inexpensive in construction, and non-proprietary (thus permitting in-house designs involving sensitive process technology).

Valve Trays

One shortcoming of the sieve tray is that its "turndown ratio" (ability to operate over a broad range of vapor rates) is limited, especially for low pressure service where pressure drop restrictions require high open areas for vapor flow. The resultant weeping of liquid, at low vapor rates, impairs efficiency. While the bubble-cap tray can meet the usual turndown requirements, its complicated geometry and its high cost has placed it at a disadvantage, as discussed earlier. An intermediate-cost alternative was developed early in the 1950-1975 period as the variable-orifice valve tray. In such a device, a liftable valve can change position as vapor flow varies, and thus can provide a changing orifice area to give a constant pressure drop. Although the Coffey still contained some variable-orifice devices, and while some field experimentation had been carried out with a so-called "rivet tray" before World War II, the modern valve tray was first installed by I. E. Nutter in 1951, based on research that had begun in 1945 (52). The tray had rectangular valves, weighted such that one side of the valve would lift first as vapor rate increased; the assemblage was called a float valve tray. Its descendant is available today from Nutter Engineering Company. The Flexitray of Koch Engineering Company followed in 1953; for this tray the valves were round and were retained by four-posted spiders (72). A third manufacturer, Glitsch, Inc., soon introduced the Ballast tray which contained specially-weighted dual valves specifically designed for the fluid densities and rates under consideration.

These initial valve trays were, of course, expensive to fabricate compared with the much simpler sieve trays. Development work continued and the valve assemblies were simplified,

resulting finally in what amounts to a cross-flow sieve tray with very large perforations, each containing a simple liftable valve with integral legs to prevent its being blown out of the perforation. In some cases it has been advantageous to use a combination of small perforations, without valves, plus conventional valves.

Details on valve tray manufacture and design are available from the three vendors noted above. Information on FRI tests of the trays are available from the same sources. A number of works have been published on valve tray performance and design; representative ones are the paper by Bolles (6) and a section of the book by Billet (3). A typical valve tray is shown in Figure 15.

Packed Columns

During this time period the packed column went through a significant renaissance, with resulting greatly expanded applications. Much of the development work was done under the sponsorship of U. S. Stoneware Co., later to become the Chemical Process Products Division of Norton Co. The first break with tradition was the introduction by U. S. Stoneware of a modified Berl saddle called the ceramic Intalox saddle (40, 41) and the publication of a "how-to-do-it" manual for packed column design by the same company in 1951 (39). The Intalox saddle was designed to prevent nesting problems that had been encountered with Berl saddles. Some pioneering studies at Clarkson College by Shulman and coworkers in 1955 (64-66) did much to elucidate the hydraulic and mass transfer parameters of packed beds, and in 1959 Cornell et al. (17) published the first general correlation of packed column mass transfer based on the two-resistance theory. Notably, this correlation covered only Raschig rings and Berl saddles, other packings not being sufficiently documented to permit their inclusion.

These traditional packing shapes involved flow of vapor and liquid around them, depending on their orientation in the bed. A major breakthrough came with the importation of the pall ring from Germany by U. S. Stoneware in 1958. This ring had been developed by the BASF company; in a ceramic material it had not tested particularly well (75) but as an import it was modified to a thin-walled metal form. This packing could be called a slotted metal Raschig ring; importantly, it permitted some of the vapor and liquid to flow through the element, thus greatly increasing mass transfer efficiency as well as flow capcity while eliminating much of the form drag pressure drop associated with the earlier ring and saddle type packings. In 1961 and 1963 Eckert presented general data on the pall ring (21, 22) but it was not until the disclosure of FRI and BASF tests in 1967 and 1969 by Billet and coworkers (2, 4) that full confidence in the use of pall rings was achieved by designers. The pall ring, and its later competitors Flexiring (Koch Engineering) and Ballast ring (Glitsch, inc.) provided a new dimension for packed column technology. Of particular interest was the low pressure drop per theoretical stage that could be obtained with pall rings; as an example, Fair (24) showed that for vacuum service a 2-inch metal pall ring would give 25% or less pressure drop per theoretical stage compared with a standard valve tray. A disadvantage of pall rings was their relatively high cost, but this changed somewhat with the increased competition in the 1970's. Typical random-type packings are shown in Figure 16.

The final development of the era was that of structured, or ordered, packings (as opposed to the randomly installed elements considered earlier). These packings were specially-shaped to fit the column diameter and bed height requirements and because of their high void fractions were known to require careful distribution of reflux and feed liquids. An early version of ordered packing was panapak, disclosed by Scofield in 1950 (62a). For small columns various types of knitted mesh packings had been developed. In the mid-1960's a special packing made from woven wire mesh and arranged in rows of vertical, corrugated elements was introduced by Sulzer Brothers in Switzerland. In 1969 Billet et al.(4) published test data on the Sulzer BX material, and other papers followed, originating with Sulzer or with the United States licensee, Koch Engineering Co.

The Sulzer packing gave excellent performance, but was quite expensive and its use was generally limited to smaller-scale specialty distillations of high-valued materials. To combat this cost disadvantage, Sulzer developed a similar packing made from sheet metal, called Mellapak (49) and marketed in the United States as Koch Flexipac. The Sulzer and Flexipac materials appeared to give even lower pressure drops per theoretical stage than pall rings, and their application for larger scale vacuum columns became more attractive with the advent of the lower-cost Flexipac. As the 1950-1975 period ended, the use of packings in columns had become well established and the ordered packings were very much in the spotlight. A section of Flexipac is shown in Figure 17.

Other Devices

Various other devices were introduced during this time period, but had relatively short lives, or have remained useful for a few special applications. These include the Koch Kascade tray (12), the Koch Benturi tray (71), The Leva film tray (42), and variations on the Kittle tray, mentioned earlier. In 1971 Nygren and Connolly(52a) summarized information on a number of distillation column devices, primarily those used in vacuum column service; the devices included the Eckey horizontal column, Glitsch-grid packing, Kloss and neo-Kloss packings, and special wiped-film contactors for handling high-boiling and viscous liquids. As noted, some of these devices are available today, and enjoy reasonable sales for special purposes.

PRESENT AND FUTURE

During the past few years there has been increasing interest in the high-performance packings of the through-flow type. It has been recognized that these packings require careful initial distribution of liquid since, unlike the situation with the more traditional "around-flow" packings, the bed itself cannot be depended upon to redistribute the liquid. Thus, much of the current work, for example at FRI, deals with distributor design.

New packings are being introduced. One of these, metal Intalox saddles (69), appears to outperform pall rings and to compete with ordered packings. A metal Intalox saddle is shown in Figure 16. Glitsch, Inc. has introduced an ordered packing called Gempak (13) which in general appearance resembles the Koch/Sulzer Flexipac. A section of Gempak is shown in Figure 17. Other packings, such as the ordered material called Montz, marketed by Chem-Pro Equipment Co., are entering the the contest. It appears that competition will be lively in the packing business, with cost/performance criteria prevailing, as usual. Word has been received that some very large vacuum columns are being changed out from trays to the high-performance packings.

Tray devices have not been completely neglected. The Parastillation tray has been described recently by Canfield (8a); this is a half-tray of the sieve type, and is inserted in the column on either side of a center baffle. Theoretical calculations, as well as tests at the University of Manchester, show that two half-trays are equivalent to about 1.3 conventional trays with the same spacing.

Work continues on the study of mass

transfer mechanisms in tray and packed columns. Commercial-scale data banks have been assembled (7, 11) to enable validation of mass transfer models against actual performance. Much of the tray work deals with an attempt to understand the two-phase contacting regimes in which a majority of the mass transfer is presumed to occur. Most of the fundamental work is being carried out in Europe, but a certain amount is continuing at Fractionation Research, Inc., and in some of the United States universities.

By and large, tray colums are being designed by computer routines, and emphasis is being placed on optimization of the designs. Computer calculation of theoretical stages has been almost routine for a number of years.

Although current interest in packed columns is quite great, the tray columns continue to be of interest for atmospheric and pressure type services. They have the advantage of better documentation and more reliable design procedures, and often are more economical on a total cost basis.

As we move through the 1980's, with full recognition of the energy intensiveness of distillation, we can expect to see relatively little displacement of distillation by alternate separation methods, at least for the larger scale process throughputs. Thus, development of distillation devices will continue. The result will be improved separation efficiency at lower pressure drop and lower cost. The lively continued competition among the manufacturers of column contacting devices can only result in benefit to the user.

LITERATURE CITED

1. AIChE [American Institute of Chemical Engineers], Bubble-Tray Design Manual, New York, 1958.

2. Billet, R., Chem. Eng. Progr. 63 (9) 53 (1967).

3. Billet, R., Distillation Engineering, Chemical Publ. Co., New York (1979).

4. Billet, R., Conrad, S., Grubb, C. M., I. Chem. E. Symp. Ser. No. 32, 5-111(1969).

5. Bolles, W. L., Petrol. Proc. 11, (2) 64, (3) 82, (4) 72, (5) 109 (1956).

6. Bolles, W. L., Chem. Eng. Progr. 72 (9) 43 (1976).

7. Bolles, W. L., Fair, J. R., Chem. Eng. 89 (14) 109 (July 12, 1982).

8. Bowles, V. O., Chem. Eng. 61 (5) 174(1954).

8a. Canfield, F., Chem. Eng. Progr. 80 (2) (1984).

9. Carey, J. S., Section 12 in Chemical Engineers' Handbook, J. H. Perry (Editor), 1st ed., McGraw-Hill Book Co., New York (1934).

10. Cellier-Blumenthal, J. B., French patents in 1813, 1818 and 1834.

11. Chan, H., Fair, J. R., Ind. Eng. Chem. Proc. Des. Devel., in press.

12. Chemical Engineering 61 (5)124(May 1954).

13. Chen, G. K., Kitterman, L., Shieh, J., Chem. Eng. Progr. 79 (9) 46 (1983).

14. Cicalese, J. J., Davies, J. A., Harrington, P. J., Houghland, G. S., Hutchinson, A. J. L., Walsh, T. J., Petrol. Refiner 26, 431, 495 (1947). [Presented at 26th Annual Meeting, Division of Refining, American Petroleum Institute, Chicago, November 1946.]

15. Coffey, A., British Patent 5974 (1830).

16. Colburn, A. P., Ind. Eng. Chem. 28, 526 (1936).

17. Cornell, D., Knapp, W. G., Fair, J. R., Chem. Eng. Progr. 56 (7) 68 (1960).

18. Dauphiné, T. C., Sc. D. Thesis, Massachusetts Institute of Technology, 1939.

19. Davies, J. A., Ind. Eng. Chem. 39, 774 (1947).

20. Delnicki, W. V., Wagner, J. L., Chem. Eng. Progr. 66 (3) 50 (1970).

21. Eckert, J. S., Chem. Eng. Progr. 57 (9) 54 (1961).

22. Eckert, J. S., Chem. Eng. Progr. 59 (5) 76 (1963).

23. Fair, J. R., Chapter 15 in Design of Equilibrium Stage Processes by B.D. Smith, McGraw-Hill Book Co., New York (1963).

24. Fair, J. R., Chem. Eng. Progr. 66 (3) 45 (1970).

25. Forbes, R. J., Short History of the Art of Distillation, E. J. Brill, Leiden (1948).

26. Forgrieve, J., *International Symposium on Distillation*, 185, Instn. Chem. Engrs., London (1960).

27. Frank, J. C., Geyer, G. R., Kehde, H., *Chem. Eng. Progr.* 65 (2) 79 (1969).

28. Gerster, J. A., Colburn, A. P., Bonnet, W., Carmody, T. W., *Chem. Eng. Progr.* 45 (12) 716 (1949).

29. Good, A. J., Hutchinson, M. H., Rousseau, W. C., *Ind. Eng. Chem.* 34, 1445 (1942).

30. Harrington, P. J., Bragg, B. L., Rhys, C., *Petrol. Refiner* 24, 502 (1945).

31. Hausbrand, E., *Rektifizier-und Destillier-Apparate*, 3rd ed., Julius Springer, Berlin (1916).

32. Hughmark, G. A., O'Connell, H. E., *Chem. Eng. Progr.* 53, 127 (1957).

33. Hutchinson, M. H., Baddour, R. F., *Chem. Eng. Progr.* 52, 503 (1956).

34. Hutchinson, M. H., Buron, A. G., Miller, B. P., paper presented at Los Angeles AIChE meeting, 1949.

35. Jones, J. B., Pyle, C., *Chem. Eng. Progr.* 51, 424 (1955).

36. Kastanek, F., Standart, G., *Sepn. Sci.* 2, 439 (1967).

37. Kirschbaum, E., *Destillier-und Rektifiziertechnik*, 3rd ed., Springer Verlag, Berlin (1960). [See also 4th ed., 1969]

38. Leslie, E. H., Coats, H. B., Chapter 19 in *Twenty-Five Years of Chemical Engineering Progress, 1908-1933*, AIChE, New York (1933).

39. Leva, M., *Tower Packings and Packed Tower Design*, U. S. Stoneware Co., Akron (1951).

40. Leva, M., U. S. Patent 2,639,909.

41. Leva, M., *AIChE J.* 1, 224 (1956).

42. Leva, M., *Chem. Eng. Progr.* 67 (3) 65 (1970).

43. Lewis, W. K., *Ind. Eng. Chem.* 14, 492 (1922).

44. Lewis, W. K., Matheson, G. L., *Ind. Eng. Chem.* 24, 494 (1932).

45. Lewis, W. K., *Ind. Eng. Chem.* 28, 399 (1936).

46. Maerker, M., Delbrueck, H., *Spiritus Fabrikation*, 813, Berlin (1908).

47. Mayfield, F. D., Church, W. L., Green, A. C., Lee, D. C., Rasmussen, R. W., *Ind. Eng. Chem.* 44, 2238 (1952).

48. McCabe, W. L., Thiele, E. W., *Ind. Eng. Chem.* 17, 605 (1925).

49. Meier, W., Stoecker, W. D., Weinstein, B., *Chem. Eng. Progr.* 83 (11) 71 (1977)

50. Muller, H. M., Othmer, D. F., *Ind. Eng. Chem.* 51, 625 (1959).

51. Murphree, E. V., *Ind. Eng. Chem.* 17, 747 (1925).

52. Nutter, I. E., *Chem. Eng.* 61 (5) 176 (1954).

52a. Nygren, P. G., Connolly, G. K. S., *Chem. Eng. Progr.* 67 (3) 49 (1971).

53. Perrier, A., British Patent 4694 (1822).

54. Peters, W. A., *Ind. Eng. Chem.* 14, 476 (1922).

55. Pfeiffenberger, C., *Chem. Week* 72, 63 (April 4, 1953).

56. Phillips, British Patent 11,965 (1847).

57. Raichle, L., Billet, R., *Chemie-Ing-Techn.* 35, 831 (1963).

58. Robinson, C. S., *Elements of Fractional Distillation*, McGraw-Hill Book Co., New York (1922). [Revised 1930 and 1939]

59. Robinson, C. S., Gilliland, E. R., *Elements of Fractional Distillation*, McGraw-Hill Book Co., New York (1950).

60. Rush, F. E., Stirba, C., *AIChE J.* 3, 336 (1957).

61. Sakata, M., Yanagi, T., *I. Chem. E. Symp. Ser. No. 56*, 3.2/21 (1979).

62. Schneible, C. B., U. S. Patents 2,596,104, -105, -106 (1952).

62a. Scofield, R. C., *Chem. Eng. Progr.* 46, 405 (1950).

63. Shell Development Co., <u>Chem</u>. <u>Eng</u>. <u>Progr</u>. <u>50</u>, 57 (1954).

64. Shulman, H. L., Ullrich, C. F., Wells, N., <u>AIChE</u> <u>J</u>. <u>1</u>, 247 (1955).

65. Shulman, H. L., Ullrich, C. F., Proulx, A. Z., Zimmerman, J. O., <u>AIChE</u> <u>J</u>. <u>1</u>, 259 (1955).

66. Shulman, H. F., Ullrich, C. F., Wells, N., Proulx, A. Z., <u>AIChE</u> <u>J</u>. <u>1</u>, 259 (1955).

67. Sorel, E., <u>Compt</u>. <u>rend</u>. 58, 1128, 1204, 1317 (1889); <u>68</u>, 1213 (1894).

68. Souders, M., Brown, G. G., <u>Ind</u>. <u>Eng</u>. <u>Chem</u>. <u>24</u>, 517 (1932).

69. Strigle, R. F., Porter, K. E., <u>I</u>. <u>Chem</u>. <u>E</u>. <u>Symp</u>. <u>Ser</u>. <u>No</u>. 56, 3.3/19 (1979).

70. Thiele, E. W., Geddes, R. L., <u>Ind</u>. <u>Eng</u>. <u>Chem</u>. <u>25</u>, 289 (1933).

71. Thornton, D. P., <u>Petrol</u>. <u>Proc</u>. <u>7</u>, 623 (1952).

72. Thrift, G. C., Chem. Eng. 61 (5) 177(1954).

73. Underwood, A. J. V., <u>Trans</u>. <u>Instn</u>. <u>Chem</u>. <u>Engrs</u>. <u>13</u>, 34 (1935).

74. Whitman, W. G., <u>Chem</u>. <u>Met</u>. <u>Eng</u>. <u>29</u>, 146 (1923).

75. Zuiderweg, F. J., Verburg, H., Gilissen, F. A. H., <u>International</u> <u>Symposium</u> <u>on</u> <u>Distillation</u>, 201, Instn. Chem. Engrs., London (1960).

76. Zuiderweg, F. J., deGroot, J. H., Meeboer, B., van der Meer, D., <u>I</u>. <u>Chem</u>. <u>E</u>. <u>Symp</u>. <u>Ser</u>. <u>No</u>. <u>32</u>, 5-78 (1969).

Figure 2. Early still, with addition of live steam for the distillation of gums and resins.

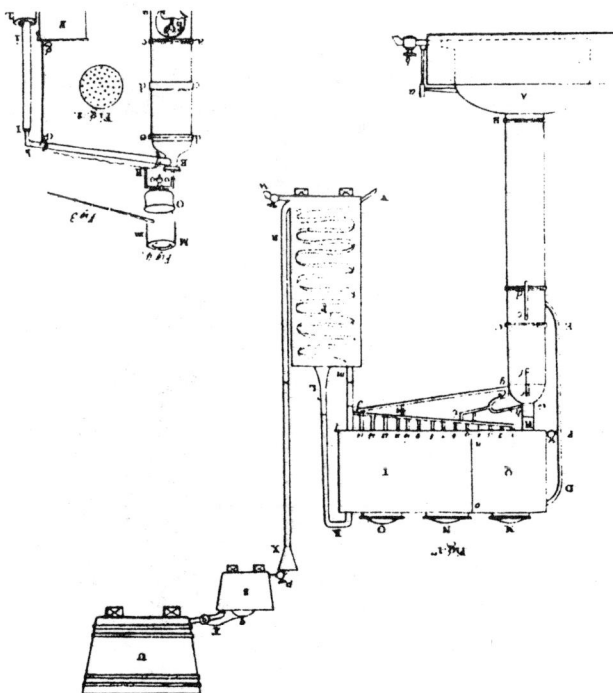

Figure 3. Cellier-Blumenthal still of 1818, with perforated plates.

Figure 1. Early still, with provision for steam contacting.

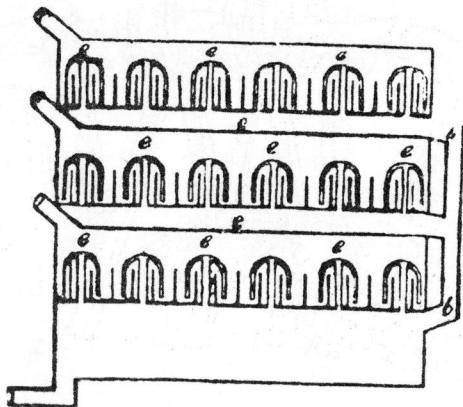

Figure 4. Bubble-cap dispersing device of Perrier (1822).

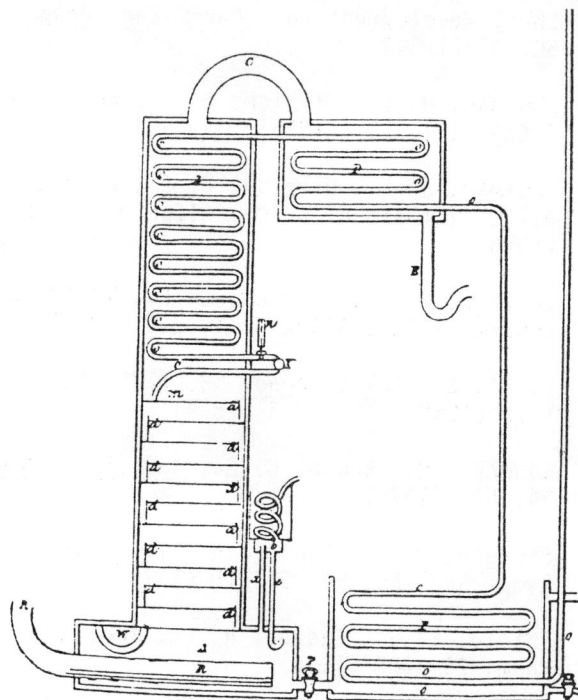

Figure 6. Sieve tray column of Coffey (1830).

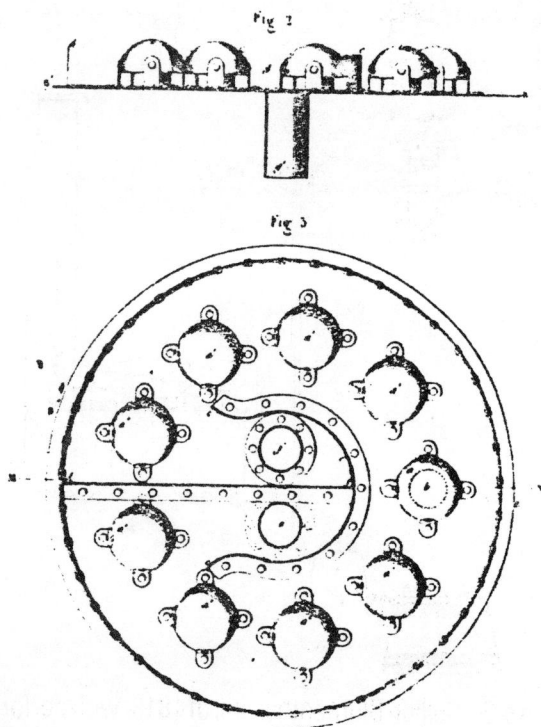

Figure 5. Cellier-Blumenthal design of bubble-cap tray (1834).

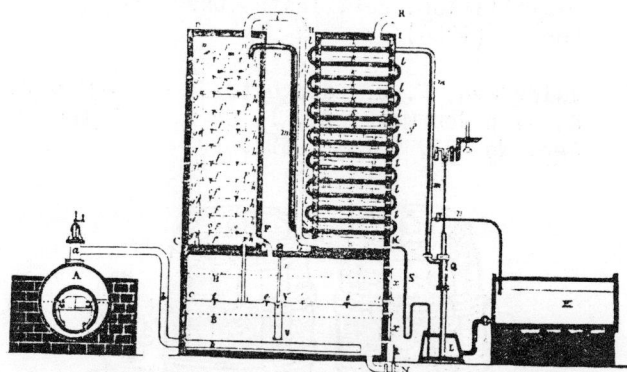

Figure 7. Later design of distillation column by Coffey.

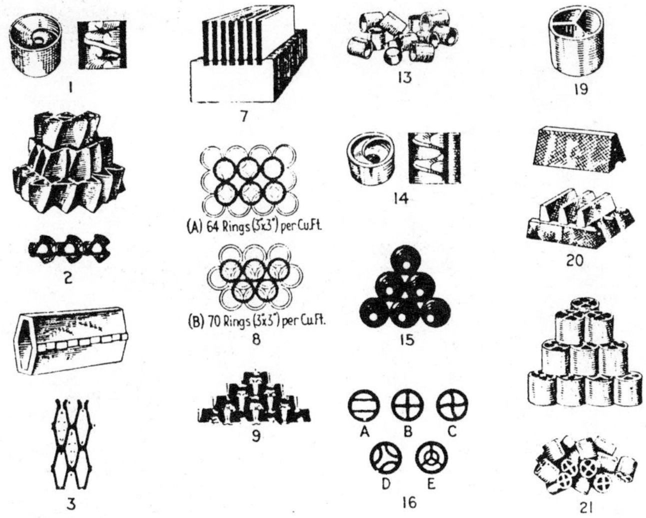

Figure 8. Examples of early packing elements (Carey,1934).

Figure 9. Various designs of slotted bubble caps (Glitsch, Inc.).

Figure 10. Two-pass bubble-cap tray with I-beam support (Glitsch, Inc.).

Figure 11. Uniflux tray of Mobil Oil Company.

Figure 12. Turbogrid tray of Shell Development Company.

Figure 14. Crossflow sieve tray (Fractionation Research, Inc.).

Figure 13. Dualflow tray (Fractionation Research, Inc.).

Figure 15. Typical valve tray (Koch Engineering Co.).

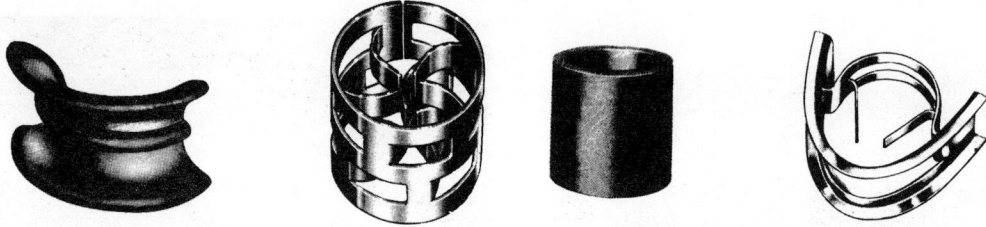

CERAMIC INTALOX SADDLE   PALL RING   RASCHIG RING   METAL INTALOX SADDLE

Figure 16. Typical dumped packings.

Figure 17. Structured packings.

HISTORY OF THE DEVELOPMENT OF DISTRILLATION COMPUTER MODELS

Charles D. Holland ■ Professor and Head, Department of Chemical Engineering, Texas A&M University

With the advent of high speed computers has come the development of new algorithms for the solution of both steady state and unsteady state distillation problems. The development has followed two avenues of approach: namely, the development of improved physical models and the application of improved mathematical techniques.

SCOPE

Although distillation was known and practiced in antiquity and a commercial still had been developed by Coffey in 1832, the theory of distillation was not studied until the work of Soral (68) in 1893. Among the other early workers were Lord Rayleigh (45), Ponchon (57) and Lewis (43). Books by Robinson (60) and Young (80) were written on the subject in 1922 and 1923. The graphical design procedures for the separation of binary mixtures was proposed by McCabe and Thiele (48) in 1925. This, procedure constituted a bench mark in the development of the theory of distillation. The next major developments were the procedures proposed involving the separation of multi-component mixtures through the use of distil-lation. These two procedures were proposed in 1932 and 1933, respectively. However, since the use of these methods was very laborious, the remainder of the 1930's, 1940's, and the early 1950's was devoted to the development of analytical solutions for simple models and the development of approximate and empirical design procedures for more complex models. Some of these methods have been described by Robinson and Gilliland (61); see also References (1,52,78).

With the advent of computers attention was redirected to the methods proposed by Lewis and Matheson (44) and by Thiele and Geddes (71). The fundamental difference between these two methods lies in the choice of the independent variables. In the Lewis and Matheson method the product distributions were taken to be the independent variables and in the Thiele and Geddes method the temperatures were taken to be the independent variables. In essentially all of the methods in use today, the temperatures are either selected as the independent variables or included in the set of independent variables.

A chronological development of the computer models used for making steady state and unsteady distillation calculations follows.

The Lewis and Matheson Method (44)

The sketch of a distillation column with the numbering system for the stages and the notation of the streams is displayed in Figure 1. The stages are numbered from the top down; the accumulator is assigned the number j = 1, the feed plate the number f, and the reboiler the number N. In this method the top of the column and each stage above the feed plate is enclosed by the material balance,

$$y_{j+1,i} = \frac{L_j x_{ji}}{V_{j+1}} + \frac{DX_{Di}}{V_{j+1}} \quad \begin{matrix}(j=1,2,\ldots,f-1)\\(i=1,2,111,c)\end{matrix} \quad (1)$$

where i is the counting integer for the components and j is the counting integer for the stages. Below the feed plate, the bottom of the column and each stage is enclosed by the component-material balance,

15

$$x_{j-1,i} = \frac{V_j y_{ji}}{L_{j-1}} + \frac{Bx_{Bi}}{L_{j-1}} \tag{2}$$

On the basis of sets of assumed values for $\{DX_{Di}\}$ and $\{Bx_{Bi}\}$, sets of independent calculations are made from each end of the column to the feed plate. The calculations consist of the alternate use of the component-material balances and equilibrium relationships. For a column having a total condenser, $x_{1i} = X_{Di}$, and Equation (1) gives $y_{2i} = X_{Di}$ for any set of assumed values for the total flow rates. A dew point calculation on the $\{y_{2i}\}$ gives the $\{x_{2i}\}$. Substitution of the $\{x_{2i}\}$ into the Equation (1) gives the set $\{y_{3i}\}$. This procedure is continued to the feed plate with the calculation of the $\{y_{fi}\}$. In a similar manner, one begins at the bottom of the column and calculates sets of liquid compositions $\{x_{ji}\}$ by use of Equation (2). Then by making a bubble point calculation on the $\{x_{ji}\}$, the corresponding set of $\{y_{ji}\}$ is obtained. Continuation of this procedure yields the composition of the vapor leaving the feed plate, the $\{y_{fi}\}$. If this set of composition "matches" [the corresponding y_{fi}'s of the two sets are equal], then the correct sets $\{DX_{Di}\}$ and $\{Bx_{Bi}\}$ were assumed. In order to justify the use of the set of total flow rates, it is necessary to recalculate the total flow rates (on the basis of the most recent values of all other variables) by use of energy balances. A θ method of convergence for making improved choices of the product distributions $\{DX_{Di}\}$ and $\{Bx_{Bi}\}$ between successive trials was proposed by Lyster et al. (46) in 1959. A variation of this version of the θ method was later proposed by Peiser (56).

The Mismatch Convergence Method

One of the first methods to be proposed for use on computers was based on the Lewis and Matheson method. In conjunction with the Lewis and Matheson method, a procedure for improving the next guess for the $\{DX_{Di}\}$ and $\{Bx_{Bi}\}$ was used. This procedure was based on the mismatch at the feed plate. This method was developed in 1953 by Mr. W. M. Harp [Exxon, Retired] and later tested and modified by Bonner (9).

The Thiele and Geddes Method (71)

In this method, the temperatures and total flow rates are taken to be the independent variables. For example when each component-material balance encloses only one stage, the complete set of component-material balances may

be stated in the following matrix form:

$$\underline{A}_i \underline{v}_i = -\underline{\delta}_i \quad (i=1,2,\ldots,c) \tag{3}$$

where

$$\underline{A}_i = \begin{bmatrix} -\rho_{1i} & 1 & 0 & & 0 & 0 \\ A_{1i} & -\rho_{2i} & 1 & & 0 & 0 \\ \cdots & \cdots & \cdots & \cdots & \cdots & \cdots \\ 0 & 0 & & A_{N-2,i} & -\rho_{N-1,i} & 1 \\ 0 & 0 & & 0 & A_{N-1,i} & -\rho_{Ni} \end{bmatrix}$$

$$\underline{v}_i = [d_i \; v_{2i} \; v_{3i} \; \cdots \; v_{Ni}]^T$$

$$\underline{\delta}_i = [0 \; \cdots \; 0 \; v_{Fi} \; \ell_{Fi} \; 0 \; \cdots \; 0]^T$$

$\rho_{ji} = 1 + A_{ji}; \quad A_{ji} = L_j/K_{ji}V_j \;\; (j \neq 1, N);$

$A_{1i} = L_1/K_{1i}D$ (for a partial condenser)

$A_{1i} = L_1/D$ (for a total condenser)

$A_{Ni} = B/K_{Ni}V_N$

A partial condenser is one in which all of the distillate product is withdrawn as a vapor, and a total condenser is one in which all of the distillate is withdrawn as a liquid.

The elements v_{Fi} and ℓ_{Fi} of the vector $\underline{\delta}_i$ are the vapor and liquid rates of component i in a partially vaporized feed. For bubble-point liquid and subcooled feeds, $v_{Fi} = 0$, $\ell_{Fi} = FX_i$. For a dew-point vapor and superheated feeds, $\ell_{Fi} = 0$, $v_{Fi} = FX_i$.

Examination of Equation (1) shows that for assumed sets $\{T_j\}$ and $\{L_j\}$, and a specified pressure P, the elements of $\underline{A}_i$ may be evaluated for each component i. Thus, the first step of the calculational procedure may be carried out; that is, Equation (3) may be solved for $\underline{v}_i$ for each component i. Since $\underline{A}_i$ is tridiagonal, $\underline{v}_i$ is usually found by use of recurrence formulas such as the Thomas algorithm or variations of it (10,31).

THE θ METHOD OF CONVERGENCE

The θ method of convergence consists of a method for finding improved sets of mole fractions and the corresponding temperatures (46). After the component-material balances have been solved as described above, it will generally be found that the sum of the calculated values of d_i [denoted by $(d_i)_{ca}$] do not equal to the specified value of D. The θ method consists of a procedure for finding an improved set of d_i's and b_i's (called "corrected" values) which have the correct sum

$$\sum_{i=1}^{c} d_i = D \qquad (4)$$

and which are in overall component-material balance

$$FX_i = b_i + d_i \ , \ (i=1,2,\ldots,c) \qquad (5)$$

The minimum number of parameters required to satisfy Equations (4) and (5) is one. Let this parameter be denoted by θ and defined by

$$\frac{b_i}{d_i} = \theta \left(\frac{b_i}{d_i}\right)_{ca} \quad (i=1,2,\ldots,c) \qquad (6)$$

where θ is of course greater than zero. Elimination of b_i from Equations (5) and (6) gives

$$d_i = \frac{FX_i}{1 + \theta\left(\frac{b_i}{d_i}\right)} = \frac{(d_i)_{ca} FX_i}{(d_i)_{ca} + \theta(b_i)_{ca}} \qquad (7)$$

To avoid numerical difficulties which arise when $(d_i)_{ca} = 0$, the new variables p_i is introduced. Its definition is

$$p_i = \frac{FX_i}{(d_i)_{ca} + \theta(b_i)_{ca}} \qquad (8)$$

Then

$$d_i = p_i(d_i)_{ca} \qquad (9)$$

Thus, the desired value of θ may be found by restating Equation (4) in functional form and replacing d_i by its equivalent [Equation (9)] to give

$$g(\theta) = \sum_{i=1}^{c} p_i(d_i)_{ca} - D \qquad (10)$$

The desired value of θ is readily found by use of Newton's method.

CORRECTED MOLE FRACTIONS AND TEMPERATURES

The set of corrected d_i's found by the θ method is used to compute new sets of compo-

sitions as follows:

$$x_{ji} = \frac{(\ell_{ji})_{ca}P_i}{\sum_{i=1}^{c}(\ell_{ji})_{ca}P_i} \ , \ y_{ji} = \frac{(v_{ji})_{ca}P_i}{\sum_{i=1}^{c}(v_{ji})_{ca}P_i} \qquad (11)$$

These compositions are used in the K_b method for finding a new set of temperatures. In this method, the $\{x_{ji}\}$ or $\{y_{ji}\}$ found by use of Equation (11) may be used to compute K_{jb} by use of either one of the following equivalent expressions

$$K_{jb} = \frac{1}{\sum_{i=1}^{c}\alpha_{ji}x_{ji}} \qquad (12)$$

$$K_{jb} = \sum_{i=1}^{c} y_{ji}/\alpha_{ji} \qquad (13)$$

In the application of Equations (12) and (13), the α_{ji}'s are evaluated at the same set of T_j's used to solved the component-material balances [Equation (3)], and the compositions to be used are those given by Equation (11). After a number value has been obtained for K_{jb} by use of either Equation (12) or (13), the expression for K_{jb} as a function of T is solved for the temperature.

Since the only restriction on the picking of K_b is that $K_b \neq 0$, a function may be selected which is both easily solved for temperatures and which promotes convergence. One such function which has been used to solve a large number of numerical examples is of the form

$$\log_e K_{jb} = \frac{a}{T_j} + b \qquad (14)$$

where the constants a and b are evaluated on the basis of the values of K at the upper and lower limits of the curve fits of the mid-boiling component of the mixture or one just lighter.

The corrected compositions and the new temperatures are used in the enthalpy balances to determine a new set of total flow rates for the next trial.

ENTHALPY BALANCES

In order to promote stability in the solution of problems involving wide-boiling mixtures, the constant-composition form of the enthalpy balances has been recommended by Holland et al. (31,36). The equations for this method are obtained by using the component material balances to eliminate one of the total flow rates from the enthalpy balance enclosing each stage. For any stage j above plate f-1 in the rectifying section, the expression for computing the total flow rate of the liquid is given by

$$L_j = \frac{Q_C - D \sum\limits_{i=1}^{c} (H_{j+1,i} - H_{Di}) X_{Di}}{\sum\limits_{i=1}^{c} (H_{j+1,i} - h_{ji}) x_{ji}} \qquad (15)$$

Extensions of this method for handling complex columns and systems of interconnected columns have been proposed, developed, and demonstrated (30,31,36). The performance of this method is demonstrated by use of Example 1; see Tables 1, 2, 3, and 4.

Relaxation Methods

Another early calculational procedure proposed by Rose et al. (63) in 1958 for finding numerical solutions was the relaxation method. In this method, the calculational procedure is initiated at some known condition which one might encounter in practice such as a column full of the liquid feed at say its bubble point temperature at the column pressure and to follow it to steady state operation by finding the successive solutions of the unsteady state equations. For any stage j (j≠1,f-1,f,N), the dynamic form of the component material balance is given by

$$V_{j+1}y_{j+1,i} + L_{j-1}x_{j-1,i} - V_j y_{ji} - L_j x_{ji}$$

$$= U_j \frac{dx_{ji}}{dt} \qquad (16)$$

where it is supposed that the holdup of the vapor is negligible and the holdup U_j of the liquid is independent of time. This equation was solved by use of Euler's method,

$$x_{ji}\Big|_{t_{n+1}} = x_{ji}\Big|_{t_n} + \Delta t \frac{dx_{ji}}{dt}\Big|_{t_n} \qquad (17)$$

After the derivative in Equation (16) has been replaced by Euler's formula [Equation (17)], an expression for computing the numerical value of x_{ji} at time t_{n+1} is obtained. A

bubble point calculation on these values of x_{ji} gives the y_{ji}'s at time t_{n+1}. After this process has been repeated for each stage, it is then repeated for the next time step.

Ball (6,36) suggested an improvement of the relaxation method of Rose wherein the differential equations were solved by use of the two-point implicit method instead of Euler's method. In the two point implicit method, the value of x_{ji} at time t_{n+1} is given by

$$x_{ji}\Big|_{t_{n+1}} = x_{ji}\Big|_{t_n} + \Delta t \left[\beta \frac{dx_{ji}}{dt}\Big|_{t_{n+1}} + (1-\beta) \frac{dx_{ji}}{dt}\Big|_{t_n} \right] \qquad (18)$$

where $0 \le \beta \le 1$. To achieve rapid convergence, Ball recommended values of $\beta > 1/2$, and in fact he generally used the pure implicit form obtained by setting $\beta = 1$. In addition, he applied the θ method at the end of each time step to place the column in component-material balance and in agreement with the specified value of the distillate rate D.

The Method of Greenstadt et al. (25): An Application of the Newton-Raphson Method

Although a variety of workers, as discussed below, have proposed the use of the Newton-Raphson method for the solution of the nonlinear equations describing a distillation column, it appears that Greenstadt et al. (25) were the first to use it to obtain the solution to these equations by use of a computer.

In the Newton-Raphson method, each multivariable function is approximated by use of the linear terms of a Taylor series expansion. Consider for example the functions $f_1, f_2, \ldots f_n$ of the independent variables $x_1, x_2, \ldots, x_n$. Then,

$$f_1(\underline{x}) = f_1(\underline{x}_0) + \frac{\partial f_1}{\partial x_1} \Delta x_1 + \frac{\partial f_1}{\partial x_2} \Delta x_2$$

$$+ \ldots + \frac{\partial f_1}{\partial x_n} \Delta x_n$$

$$f_2(\underline{x}) = f_2(\underline{x}_0) + \frac{\partial f_2}{\partial x_1} \Delta x_1 + \frac{\partial f_2}{\partial x_2} \Delta x_2 + \cdots$$

$$+ \frac{\partial f_2}{\partial x_n} \Delta x_n$$

$$f_n(\underline{x}) = f_n(\underline{x}_0) + \frac{\partial f_n}{\partial x_1} \Delta x_1 + \frac{\partial f_n}{\partial x_2} \Delta x_2 + \cdots$$

$$+ \frac{\partial f_n}{\partial x_n} \tag{19}$$

where all of the derivatives are evaluated at the set of variables

$$\underline{x}_0 = [x_{1,0} \ x_{2,0} \ \cdots \ x_{n,0}]^T$$

Let $\underline{x}_0$ denote the set of assumed values of the variables. Now let $\underline{x}$ be selected such that each function $f_j(\underline{x})$ [$j = 1,2,\ldots,n$] appearing in Equation (18) is equal to zero. Let this value of $\underline{x}$ be denoted by $\underline{x}_1$. Then Equation (18) reduces to a set of linear equations in the variables Δx_1, Δx_2, ..., Δx_n. These equations have the following matrix representation for the kth iteration.

$$\underline{J}_k \underline{\Delta x}_k = -\underline{f}_k \tag{20}$$

where

$$\underline{J}_k = \begin{bmatrix} \frac{\partial f_1}{\partial x_1} & \cdots & \frac{\partial f_1}{\partial x_n} \\ \cdot & & \cdot \\ \cdot & & \cdot \\ \cdot & & \cdot \\ \frac{\partial f_n}{\partial x_1} & \cdots & \frac{\partial f_n}{\partial x_n} \end{bmatrix}$$

$$\underline{\Delta x}_k = \underline{x}_{k+1} - \underline{x}_k = [\Delta x_1 \ \Delta x_2 \ \cdots \ \Delta x_n]^T$$

$$\underline{f}_k = [f_1 f_2 \ \cdots \ f_n]^T$$

After Eqaution (20) has been solved for $\underline{\Delta x}_k$ on the basis of the assumed values $\underline{x}_k$, the values of the variables to be assumed for the next trial are then found as follows:

$$x_{1,k+1} = x_{1,k} + \Delta x_{1,k}$$
$$\cdot \qquad \cdot \qquad \cdot$$
$$\cdot \qquad \cdot \qquad \cdot \tag{21}$$
$$\cdot \qquad \cdot \qquad \cdot$$
$$x_{n,k+1} = x_{n,k} + \Delta x_{n,k}$$

The process is repeated until the absolute value of each function has been reduced to a value less than some small preassigned number.

In the calculational procedure proposed by Greenstadt. et al. (25) the equations describing each stage of the column were solved sequentially, first for the rectifying section and then for the stripping section.

In that the calculational procedure was initiated on the basis of assumed values for the product distributions $\{d_i\}$ and $\{b_i\}$, it is like the Lewis and Matheson method (44). It differed from the Lewis and Matheson method in that instead of making bubble point calculations, Greenstadt et al. solved these equations simultaneously with the remaining equations for the stage.

The Newton-Raphson formulation for any interior stage j consisted of (2c+3) independent variables $\underline{x}$. In particular the independent functions were based on the following (2c+3) independent equations:

(1) the c equilibrium relationships,

$$y_{ji} = K_{ji} x_{ji}$$

(2) the c component-material balances

(3) the two summation equations,

$$\sum_{i=1}^{c} x_{ji} = 1, \quad \sum_{i=1}^{c} y_{ji} = 1$$

(4) one energy balance

The 2c+3 independent variables were taken to be

$$\underline{x} = [x_{j,1} \ \cdots \ x_{j,c} y_{j,1} \ \cdots \ y_{j,c} T_j \ V_j \ L_J]^T \tag{22}$$

N and 2N Newton-Raphson Methods

Since the proposed application of the Newton-Raphson method by Greenstadt et al.

(25), a host of workers have proposed applications of this method (15,16,28,29,34,69,72, 73,76). In the 50's, 60's and early 70's the lack of computer size and storage gave rise to the N and 2N formulations of the Newton-Raphson method. In order to reduce the size of the Jacobian from say an N(2c+3) square to an N or 2N square matrix, the number of independent variables were reduced fron N(2c+3) to N or 2N.

In 1961, Sujata (69) proposed that the N temperatures be regarded as the independent variables and that they be determined by use of the N enthalpy balances. In 1963, Newman proposed that the total flow rates $\{L_j\}$ be regarded as the independent variables and the corresponding sets of temperatures needed to satisfy the component-material balances and equilibrium relationships be found by use of the Newton-Raphson method. Friday and Smith (15), Boynton (13) as well as Holland et al. (33,34) proposed similar applications of the Newton-Raphson method which were of order N.

In a series of papers, Tierney and co-workers (72,73,74) proposed a formulation of the Newton-Raphson method in which the vapor rates $\{V_j\}$ and temperatures $\{T_j\}$ were taken to be the independent variables. The partial derivatives in the Newton-Raphson method were evaluated by use of the calculus of matrices (27,28,31,33). Methods proposed by Holland co-workers are summarized in Reference (31) wherein the independent variables were taken to be the flow ratios $\{L_j/V_j\}$ and the temperatures, the $\{T_j\}$. The equations were solved by three different procedures: (1) use of the calculus matrices in the evaluation of the partial derivatives, (2) use of Broyden's method (14), which was first applied to distillation problems by Tomich (76), and (3) use of the Broyden-Bennet Algorithm (31).

To demonstrate the general approach involved in these methods, the equations for the formulation of the 2N Newton-Raphson method given in Reference (31) are presented for an absorber with N stages: see Figure 2. The independent variables $\underline{x}$ are taken to be

$$\underline{x} = [\theta_1\theta_2 \cdots \theta_N \ T_1T_2 \cdots T_N]^T \qquad (23)$$

and the 2N independent functions $\underline{f}$ are taken to be

$$\underline{f} + [F_1F_2 \cdots F_N \ G_1G_2 \cdots G_N]^T \qquad (24)$$

The variables $\{\theta_j\}$ constitute a normalized form of the variables $\{L_j/V_j\}$ which are defined as follows:

$$\frac{L_j}{V_j} = \theta_j \left(\frac{L_j}{V_j}\right)_a \qquad (j=1,2,\ldots,N) \qquad (25)$$

The quantity $(L_j/V_j)_a$ is an arbitrary constant which was taken to be equal to the most recently calculated value of L_j/V_j. The set $\{F_j\}$ consists of dew point functions which were stated in the form:

$$F_j = \frac{1}{V_j} \sum_{i=1}^{c} \left(\frac{1}{K_{ji}} - 1\right) \qquad (j=1,2,\ldots,N) \qquad (26)$$

and the $\{G_j\}$ are the enthalpy balance functions which were stated in the form:

$$G_j = \frac{\sum_{i=1}^{c} (v_{ji} H_{ji} + \ell_{ji} h_{ji})}{\sum_{i=1}^{c} (v_{j+1,i}H_{j+1,i} + \ell_{j-1,i}h_{j-1,i})} - 1 \qquad (27)$$

For each choice of the independent variables $\underline{x}$, the corresponding values of the dependent variables are found by solving the component-material balances, given by Equation (3). In this case the elements of $\underline{A}_i$, $\underline{v}_i$, and $\underline{\delta}_i$ are as follows.

$$\underline{A}_i = \begin{bmatrix} -\rho_{1i} & 1 & 0 & & 0 \\ A_{1i} & -\rho_{2i} & 1 & & 0 \\ \cdots & \cdots & \cdots & \cdots & \cdots \\ 0 & 0 & A_{N-2,i} & -\rho_{N-1,i} & 1 \\ 0 & 0 & 0 & A_{N-1,i} & -\rho_{Ni} \end{bmatrix}$$

$$\underline{\delta}_i = [\ell_{0i} \ 0 \cdots 0v_{N+1,i}]^T$$

$$A_{ji} = L_j/(K_{ji}V_j) \qquad (28)$$

$$\rho_{ji} = 1 + A_{ji} .$$

The total material balances are given by

$$\underline{R} \ \underline{V} = - \underline{F} \qquad (29)$$

where

$$R = \begin{bmatrix} -(1+R_1) & 0 & 0 & 0 & 0 \\ R_1 & -(1+R_2) & 1 & 0 & 0 \\ \cdots & \cdots & \cdots & \cdots & \cdots \\ 0 & 0 & R_{N-2} & -(1+R_{N-1}) & 1 \\ 0 & 0 & 0 & R_{N-1} & -(1+R_N) \end{bmatrix}$$

$$\underline{V} = [V_1 V_2 \cdots V_N]^T$$

$$\underline{F} = [L_0\ 0 \cdots 0\ V_{N+1}]^T$$

$$R_j = \theta_j (L_j/V_j)_a .$$

For the case where the mixtures throughout the column form ideal solutions [$\gamma_{ji}^V = \gamma_{ji}^V = 1$, $K_{ji}(P_j,T_j)$, $h_{ji} = h_{ji}(P_j,T_j)$, and $A_{ji} = A_{ji}(P_j,T_j)$], the above formulation of the Newton-Raphson method constitutes an exact application of this method. Thus, the 2N Newton-Rapshon method as formulated above for ideal solutions can be expected to exhibit the quadratic convergence of the Newton-Raphson method.

Since the 2N Newton-Raphson method is not an exact application of the Newton-Raphson method when the mixtures form highly nonideal solutions at the column conditions, it may fail to converge for problems involving such solutions. Exact applications of the Newton-Raphson method have been proposed, however, for the solution of problems involving the separation of highly nonideal solutions.

A comparison of the performance of the θ method and the 2N Newton-Raphson methods is presented in Table 7.

Banded Formulations of the Newton-Raphson Method

To obtain exact expressions for the partial derivatives it is convenient to include the compositions $\{x_{ji}, y_{ji}\}$ or the component-flow rates in $\{\ell_{ji}, v_{ji}\}$ in the set of independent variables.

Beginning with the formulation of Greenstadt et al. (25) as described previously, a number of formulations have been proposed. For the solution of isothermal liquid-liquid extraction, Roche and Staffin (62) proposed an N(c+2) formulation of the Newton-Raphson method

in 1968 in which the independent variables were taken to be $\{V_j\}$, $\{L_j\}$, and $\{x_{ji}\}$.

In 1971, Naphtali and Sandholm (53) proposed a formulation of the Newton-Raphson method in which the independent variables were taken to b N(2c+1) in number, the $\{\ell_{ji}, v_{ji}\}$ and $\{T_j\}$.

In 1972, Bruno et al. (16), proposed a formualtion in terms of N(c+1) independent variables, $\{V_j\}$, $\{T_j\}$, and $\{x_{ji}\}$. In 1976, a Newton-Raphson algorithm formulated in terms of N(c+2) was proposed by Gallun et al. (21).

In 1981, Gallun and Holland (20) presented a modification of Broyden's method for the solution of systems of sparse matrices. This numerical technique was applied to the solution of an extractive distillation problem which was formulated in terms of N[(2c+1) + 2] independent variables, the $\{\ell_{ji}\}$, $\{v_{ji}\}$, $\{T_j\}$, Q_C, and Q_R. Use of this technique avoided the necessity of obtaining analytical expressions of the partial derivatives. Furthermore, as in Broyden's method, the partial derivatives of the Jacobian were evaluated numerically only one time. A similar method proposed by Schubert (66) was also demonstrated by Gallun and Holland (20, 31). Example 2 (Tables 5 and 6) was used to compare these methods.

In order to demonstrate a typical formulation where compositions or flow rates are included in the set of independent equations, the equations presented by Holland (31) for an absorber (see Figure 2) follow. In this case the independent variables $\underline{x}$ are given by

$$\underline{x}\ [(\ell_{j,1}\ell_{j,2} \cdots \ell_{j,c} v_{j,1} v_{j,2} \cdots$$
$$v_{j,c} T_j)_{j=1,N}]^T \qquad (30)$$

where the parentheses with the subscript j=1,N means that the arguments are to be repeated for j=1,2,...,N. The corresponding functions are given by

$$f_{ji} = \frac{\gamma_{ji}^L K_{ji} \ell_{ji}}{\sum_{i=1}^{c} \ell_{ji}} - \frac{\gamma_{ji}^V v_{ji}}{\sum_{i=1}^{c} v_{ji}} \quad \begin{array}{l} (j=1,2,\ldots,N) \\ (i=1,2,\ldots,c) \end{array}$$

$$(31)$$

$$m_{ji} = v_{j+1,i} + \ell_{j-1,i} - v_{ji} - \ell_{ji}$$

$$\begin{align} &(j=1,2,\ldots,N) \\ &(i=1,2,\ldots,c) \end{align} \quad (32)$$

$$G_j = \frac{\sum\limits_{i=1}^{c} [v_{ji} \hat{H}_{ji} + \ell_{ji} \hat{h}_{ji}]}{\sum\limits_{i=1}^{c} [v_{j+1,i} \hat{H}_{j+1,i} + \ell_{j-1} \hat{h}_{j-1,i}]} - 1$$

$$(j=1,2,\ldots N) \quad (33)$$

The resulting Jacobian matrices before and after transforming to triangular form are shown in Figures 3 and 4. In additions to the above procedure or solving problems involving highly nonideal solutions, an iterative procedure has been proposed by Boston and Sullivan (10,11). A summary of the recommended uses of the various calculational procedures which have been proposed are presented in Table 8.

Techniques Proposed for Solving Other Types of Problems

Two general procedures have been recommended for solving systems of columns interconnected by recycle streams, the column modular method (24,31,75) and the system modular method (38,42). In the column modular method, each column is solved sequentially by the procedure recommended for it in Table 8, and after one complete trial has been made on the system, the capital θ method is applied. Systems involving simultaneous mass and energy transfer between columns (see Figure 5) have been solved by this technique. In the system modular method, the equations for the entire system are solved simultaneously as a single set

Several procedures have been proposed for the design of distillation columns. One of the first efforts to solve design problems by use of optimization procedures was made by Srygley and Holland (70), who used the Hooke and Jeeves method (37). Sargent and Gaminibandara (65) considered the more general problem of the optimum configuration of columns needed to effect a specified separation. More recently, an iterative procedure based on the Naphtali and Sandholm (53) formulation of the Newton-Raphson method was proposed by Ricker and Grens (59) for the minimization of the number of stages required to effect a specified separation at a given reflux ratio. The utilization of Box's complex method (12) for the solution of a variety of design and

optimization problems has been demonstrated by Al-Haj-Ali and Holland (4,5,31).

The solution of extractive and azeotropic distillation problems by use of the Almost Band Algorithm has been demonstrated by Holland et al. (31). Several distillation processes referred to as extractive distillations sometimes involve simultaneous distillation and chemical reaction. A number of calculational procedures have been proposed for solving problems on this type. Suzaki et al. (67) used a procedure based on Muller's method, while Jelinek and Halavacek employed a relaxation method (41). Utilization of the θ method as well as the Newton-Raphson method was proposed in a series of papers (40,51). Nelson (55) also proposed a formulation of the Newton-Raphson method for solving distillation problems involving simultaneous distillation and chemical reaction.

Methods for solving problems involving systems which form three phase mixtures throughout the column have been proposed by Block and Hegner (8) and by Ross and Seider (64).

Dynamic Separation Problems

The recent availability of more and better computer hardware makes it practical to determine the dynamic behavior of separation processes for different control systems and varying load conditions. The utilization of the dynamic simulation make it possible to choose the best control system for the given application prior to the construction of the process equipment.

With more and better computer hardware has come new and more powerful numerical methods for solving dynamic problems. In the early work on this topic, approximate analytical solutions were obtained by use of the calculus of matrices (3,47). Euler's method was used by Rose et al. (63) in their attempt to follow the dynamic behavior of separation processes. A more successful approach (because of its inherent stability) consisted of the combination of the θ method and the two-point implicit method proposed by Waggoner and Holland (79). The contributions of a number of other workers in this field have been summarized by Holland (35).

Recently several workers have considered the solution of the equations required to described the dynamic behavior of distillation columns (7,17,19,23,24,49,50,58,67,77). Of the numerical methods proposed in recent years, the most promising for the simulation of the dynamic behavior of separation processes are perhaps Gear's method (23,24) and the semi-implicit Runge-Kutta method (17) as modified by Michelson (49,50).

GEAR'S METHOD

Gear's method is a mulitple-point, variable-time step method which may be used to solve systems of coupled differential and algebraic equations of the general form:

$$\underline{0} = \underline{f}(\underline{y}, \underline{z}, \underline{y}'\ \underline{z}') \qquad (34)$$

$$\underline{0} = \underline{g}(\underline{y}, \underline{z}) \qquad (35)$$

The algorithm of this method is for convenience presented for a relative simple system consisting of one differential equation and one algebraic equation. In the application of Gear's method, it has been found convenient to carry the information contained in the past values of the variables (y_{n-1}, y_{n-2},...) in the form of the terms of a Taylor series expansion of the function evaluated at time t_n. The two vectors containing the past information about y and z are called the Nordsieck vectors $\underline{Y}_n$ and $\underline{Z}_n$, which are defined as follows for a kth order algorithm.

$$\underline{Y}_n = \left[y_n,\ hy'_n,\ \frac{h^2}{2!} y_n^{(2)},\ \ldots,\ \frac{h^k}{k!} y_n^{(k)} \right]^T$$
$$(36)$$

$$\underline{Z}_n = \left[z_n,\ hz'_n,\ \frac{h^2}{2!} z_n^{(2)},\ \ldots,\ \frac{h^k}{k!} z_n^{(k)} \right]$$
$$(37)$$

The name "kth order algorithm" is used to reflect the fact that if the solution to a given differential equation is given by a polynomial of degree k, then the the kth order algorithm will give an exact solution. The algorithm is applied as follows:

$$\tilde{\underline{Y}}_n = \underline{D}\ \underline{Y}_{n-1} \qquad (38)$$

$$\tilde{\underline{Z}}_n = \underline{D}\ \underline{Z}_{n-1}$$

where $\underline{D}$ is the Pascal triangle matrix. For a 3rd order algorithm,

$$\underline{D} = \begin{bmatrix} 1 & 1 & 1 & 1 \\ 0 & 1 & 2 & 3 \\ 0 & 0 & 1 & 3 \\ 0 & 0 & 0 & 1 \end{bmatrix}$$

In general, the nonzero element $d_{i+1,j+1}$ of the (j+1)st column and (i+1)st row of the Pascal triangle matrix is given by

$$d_{i+1,j+1} = \frac{j!}{(j-i)!\ i!}$$

Step 2. Find the values of b_1 and b_2 which satisfy the original equations; that is, find b_1 and b_2 such that $F_1(b_1,b_2) = F_2(b_1,b_2) = 0$ where

$$F_1(b_1,b_2) = F_1(\tilde{y}_n + \beta_{-1}\ b_1,\ h\tilde{y}'_n + b_1,$$
$$\tilde{z}_n + \beta_{-1}\ b_2,\ h\tilde{z}'_n + b_2) \qquad (40)$$

$$F_2(b_1,b_2) = F_2(\tilde{y}_n + \beta_{-1}\ b_1,\ \tilde{z}_n + \beta_{-1}\ b_2)$$
$$(41)$$

and

$$y_n = \tilde{y}_n + \beta_{-1}\ b_1,\ z_n = \tilde{z} + \beta_{-1}\ b_2 \qquad (42)$$

$$hy'_n = h\tilde{y}'_n + b_1,\ hz'_n = h\tilde{z}'_n + b_2 \qquad (43)$$

Equations (40) and (41) may be solved by the Newton-Raphson method.

Step 3. After the solution values of b_1 and b_2 have been found, the corrected values of the vectors at time t_n are computed as follows:

$$\underline{Y}_n = \tilde{\underline{Y}}_n + b_1\ \underline{L}$$
$$(44)$$
$$\underline{Z}_n = \tilde{\underline{Z}}_n + b_2\ \underline{L}$$

The parameter β_{-1} and the vector $\underline{L}$ depend upon the order of the algorithm, and they are presented in Table 9.

Example 3: The differential equation for a component-material balance for a perfect mixer at unsteady steady state operation in which the holdup U is independent of time is given by

$$FX - Lx = U \frac{dx}{dt}$$

where F is the feed rate and L is the outlet rate. For $t > 0$, $X = 0.9$, $x = 0.1$, $F = L = 100$, and $U = 50$. Use a second order Gear's method to find x at the end of the first time step h. Take $h = 0.2$. For Gear's second-order method, $\beta_{-1} = 2/3$ and $L = [2/3, 3/3, 1/3]^T$.

SOLUTION

At the above conditions, the differential equation becomes

$$x' = 1.8 - 2x$$

The elements of $\underline{Z}_0$ are $x_0 = 0.1$, and

$$hx_0' = h[1.8-2x] = 0.2[1.8-(2)(0.1)] = 0.32$$

Since

$$x^{(2)} = \frac{d[1.8-2x]}{dx} \frac{dx}{dt} = (-2) x'$$

and

$$\frac{h^2 x_0^{(2)}}{2!} = \frac{(0.2)^2}{2} (-2)(1.6) = -0.064$$

Step 1.

$$Z_1 = \underline{D}\, \underline{Z}_0 = \begin{bmatrix} 1 & 1 & 1 \\ 0 & 1 & 2 \\ 0 & 0 & 1 \end{bmatrix} \begin{bmatrix} 0.1 \\ 0.32 \\ -0.064 \end{bmatrix} = \begin{bmatrix} 0.356 \\ 0.192 \\ -0.064 \end{bmatrix}$$

Step 2.

$$x_1 = \tilde{x}_1 + \beta_{-1}\, b = 0.356 + \frac{2}{3} b$$

$$hx_1' = h\, \tilde{x}_1' + b = 0.192 + b$$

$$G(b) = h[1.8 - 2(\tilde{x}_1 + \beta_{-1}b)] - [h\, \tilde{x}_1' + b]$$

$$= (0.2) [1.8 - 2(0.356 + 2/3\, b)]$$

$$- (0.2)(0.192) - b$$

The b that makes $B(b) = 0$ is $b = 0.0202$

Step 3.

$$z_1 = z_1 + b\, \underline{L} = \begin{bmatrix} 0.356 \\ 0.192 \\ -0.064 \end{bmatrix} +$$

$$(0.0202) \begin{bmatrix} 2/3 \\ 3/3 \\ 1/3 \end{bmatrix} = \begin{bmatrix} 0.369 \\ 0.212 \\ -0.0573 \end{bmatrix}$$

Thus, the value of x at $t = 0.2$ is $x = 0.369$.

Gear's method contains a procedure for changing the size of the time step and the order of the algorithm simultaneously (23, 24). To demonstrate the application of Gear's method, Gallun et al. (22) solved a problem which involved the separation of the mixture shown in Example 2 by use of the column shown in Figure 6. For an upset consisting of a change in the set point of the temperature controller, the behavior shown in Figures 7 and 8 was predicted by the model.

SEMI-IMPLICIT RUNGE-KUTTA METHOD

The third order semi-implicit Runge-Kutta method which was originally proposed by Caillaud and Padmanabhan (17) for the solution of stiff differential equations was later modified by Michelsen (49,50). The formulas for the modified version of Michelsen follow.

$$y_{n+1} = \underline{y}_n + R_1\underline{k}_1 + R_2 k_2 + R_3\underline{k}_3 \qquad (45)$$

where

$$\underline{k}_1 = h[\underline{I} - ha\underline{J}(\underline{y}_n)]^{-1}\, \underline{f}(\underline{y}_n)$$

$$\underline{k}_2 = h[\underline{I} - ha\underline{J}(\underline{y}_n)]^{-1}\, \underline{f}(\underline{y}_n + b_2\underline{k}_1)$$

$$k_3 = [I - ha J(y_n)]^{-1}\, [b_{31}\underline{k}_1 + b_{32}\underline{k}_2)]$$

In the above expressions, $\underline{J}(\underline{y}_n)$ denotes the Jacobian matrix of the fundamental part of each differential equation of the form

$$\frac{dy}{dt} = f(t,y)$$

For a single differential equation,

$$\underline{J}(\underline{y}_n) = \left.\frac{\partial f(t,y)}{\partial y}\right|_{t_n, y_n} \qquad (46)$$

The constants in Equation (45), when evaluated to within four significant figures, have the following values

$$a = 0.4358, \quad b_2 = 3/4, \quad b_{31} = -0.6302,$$

$$b_{32} = -0.2423$$

$$R_1 = 1.038, \quad R_2 = 0.8349, \quad R_3 = 1$$

To demonstrate the application of this method, x at the end of the first time step of h = 0.2 is computed for Example 3.

$$k_1 = 0.2[1 - (0.2)(0.4358)(-2)]^{-1} (1.6)$$

$$= 0.2725$$

$$k_2 = 0.2[1 - (0.2)(0.4358)(-2)]^{-1} [1.8$$

$$- (2[0.1 + (0.75)(0.2725)])] = 0.2029$$

$$k_3 = [1 - (0.2)(0.4358)(-2)]^{-1} [(-0.6302)$$

$$(0.725) + (-0.2423)(02029)] = -0.1881$$

Thus,

$$x = 0.1 + (1.038)(0.2725) + (0.8349)$$

$$(0.2029) + (-0.1881) = 0.3642$$

The parameters listed above were selected such that the method is A-stable as discussed by Michelsen (49,50), who also modified the method such that it may be used to solve coupled differential and algebraic equations. A method for making changes in the size of the time step was also proposed by Michelsen. This procedure involved obtaining these solutions for each time step, at h and at h/2. The method is, however, explicit in the sense that no trial and error is involved.

Future Trends in Modeling

Because of the inherent energy efficiency and the sharpness of separations which can be achieved when a separation is carried out by use of a distillation process accompanied by chemical reaction, this area should prove to be a fertile field for the development of advanced modeling techniques. Also, the increased use of packing in large-diameter columns calls for additional modeling techniques for a more precise description of this separation process.

Because of the rapid rate of increase in computer hardware capability, it is anticipated that the future demand for the solution of dynamic separation problems will approach our present demand for the solution of steady state problems. Consequently, the modeling of the dynamic behavior of columns and their control systems will call for further model development in the future.

ACKNOWLEDGMENTS

This work was supported in part by Dow Chemical Company, U.S.A. and The Texas Engineering Experiment Station. This support is gratefully acknowledged.

NOTATION

a	= constant appearing in the function for K_b; see Equation (14).
A_{ji}	= absorption factor; $A_{ji} = L_j/(K_{ji}V_j)$.
$\underline{A}_i$	= square matrix appearing in the matrix equation for the component material balances; see Equation (3)
b	= constant appearing in the expression for K_b; see Equation (14).
b_i	= molar flow rate of component i in the bottom product.
b_1, b_2	= constants appearing in Gear's algorithm; see Equations (40) and (41).
B	= total molar flow rate of the bottoms.
c	= total number of components.
d_i	= molar distillate rate of component i.
D	= total molar distillate rate.
$\underline{D}$	= Pascal triangular matrix; see Equations (38) and (39).
f	= feed plate; number of feed plate.
$\underline{f}$	= vector of functions.
$\underline{f}_i$	= feed vector for component i; see Equation (3).

$\underline{F}$ = feed vector for total flow rates; see Equation (29).

F = total molar flow rate of the feed.

F_j = dew point form of the equilibrium relationship; see Equation (26).

g = g function.

G_j = enthalpy balance function; see Equations (27) and (33).

h = length of time step.

h_{ji} = enthalpy of pure component i in the liquid state at the temperature and pressure of stage j.

t = time.

t_n = time, $t = t_n$.

Δt = increment of time.

T = temperature.

U_j = total molar holdup of liquid on stage j.

v_{Fi} = molar flow rate of component i in the vapor part of the feed.

v_{ji} = molar flow rate at which component i leaves stage j.

$\underline{v}_i$ = vector of component flow rates; see Equation (3).

V_F = total molar flow rate of component i in the vapor of the feed.

V_j = total molar flow rate at which vapor leaves stage j.

$\underline{V}$ = vector of the total flow rates.

V_{N+1} = total flow rate of rich gas to an absorber.

x_{ji} = mole fraction of component i in the liquid leaving stage j.

X_i = total mole fraction of component i in the feed, regardless of state.

X_{Di} = mole fraction of component in the distillate, regardless of state.

x_{Bi} = mole fraction of component i in the bottoms product.

$\underline{\Delta x}$ = correction vector of the Newton-Raphson method; see Equation (20).

y_{ji} = mole fraction of component i in the vapor leaving stage j.

$\underline{Y}_n$ = Nordseick vector at time t_n; see Equation (36).

$\underline{\tilde{Y}}_n$ = predicted value of $\underline{Y}_n$.

$\underline{Z}_n$ = Nordseick vector at time t_n; see Equation (37).

$\underline{\tilde{Z}}_n$ = predicted value of $\underline{Z}_n$.

$\hat{h}_{ji}$ = enthalpy of component i in a liquid mixture in the liquid state at the temperature and pressure of stage j.

$\hat{H}_{ji}$ = enthalpy of pure component i in the vapor state at the temperature and pressure of stage j.

H_{ji} = enthalpy of component i in a mixture in the vapor state at the temperature and pressure of stage j.

$\underline{I}$ = identify matrix.

j = stage number.

$\underline{J}$ = Jacobian matrix; see Equation (20).

$\underline{k}$ = constants appearing in the semi-implicit Runge-Kutta method; see Equation (45).

K_{ji} = ideal solution K value; function of the temperature and pressure of stage j.

K_{jb} = base-component K value; evaluated at the temperature and pressure of stage j.

ℓ_{Fi} = molar flow rate at which component i leaves stage j in the feed.

ℓ_{ji} = molar flow rate at which component i leaves stage j in the liquid phase.

L_F = total molar flow rate of the liquid part of the feed.

L_j = total molar flow rate at which liquid leaves stage j.

P_i = dimensionless form of the distillate rate for component i; see Equations (8) and (9).

P = total pressure.

Q_C = condenser duty; net energy removed by the condenser per unit time.

Q_R = reboiler duty; net energy input to the reboiler per unit time.

$\underline{R}$ = square matrix in the total material balancer matrix equation; see Equation (29).

R_j = an element of the matrix $\underline{R}$.

GREEK LETTERS

α_{ji} = relative volatility for component i and stage j; $\alpha_{ji} = K_{ji}/K_{jb}$.

β = parameter in the two-point implicit method; see Equation (18).

γ_{ji} = activity coefficient for component i and stage j.

θ = multiplier used in θ method for single columns.

θ_j = variable proportional L_j/V_j; defined by Equation (25).

ρ_{ji} = used to represent the elements along the central diagonal of the matrix for the component material balances; see Equation (3).

REFERENCES

1. Acrivos, A. and N. R. Amundson, "On the Steady State Fractionation of Multicomponent and Complex Mixtures in an Ideal Cascade: Part 2 - The Calculation of the Minimum Reflux Ratio," Chem. Eng. Sci. 42 (No. 2), 68 (1955).

2. Amundson, N. R. and A. J. Pontinen, "Multicomponent Distillation Calculations on a Large Digital Computer," Ind. Eng. Chem. 50, No. 5, 730 (1958).

3. Acrivos, A and N. R. Amundson, "Application of Matrix Mathematics to Chemical Engineering Problems," Ind. Eng. Chem. 47, 1533 (1955).

4. Al-Haj-Ali, Najeh and C. D. Holland, "Way to Find Distillation Optimum," Hydrocarbon Process, 58, 165 (1979). Part 1 - For Conventional Column.

5. Al-Haj-Ali, Najeh and C. D. Holland, "Way to Find Distillation Optimum," Hydrocarbon Process, 58, 111 (1979).

6. Ball, W. E., "Computer Programs for Distillation," paper presented at the Machine Computation Workshop Session on Multicomponent Distillation at the 44th National Meeting, AIChE, New Orleans, February 27, (1961).

7. Ballard, D., C. Brislow and C. Kahn, "Dynamic Simulation of Multicomponent Distillation Column," paper presented at AIChE Meeting, October, (1978).

8. Block, U. and B. Hegner, "Development and Application of a Simulation Model for Three-Phase Distillation," AIChE J 22, 3 (1976).

9. Bonner, J. S., "Solution of Multicomponent Distillation Problems on a Stored Program Computer," American Petroleum Institute Quarterly, Division of Refining, (1956).

10. Boston, J. F. and S. L. Sullivan, Jr., "An Improved Algorithm for Solving Mass Balance Equations in Multistage Separation Processes," Can. J. Chem. Eng. 50, 663 (1972).

11. Boston, J. F. and S. L. Sullivan, Jr., "A New Class of Solution Methods for Multicomponent Multistage Separation Processes," Can. J. Chem. Eng. 52, 52 (1974).

12. Box, M. J., "A New Method of Constrained Optimization and a Comparison with Other Methods," Computation Journal, 8, 42 (1965).

13. Boynton, G. W., "Iteration Solves Distillation," Hydrocarbon Processing, 49, 153 (1970).

14. Broyden, C. G., "A Class of Methods for Solving Nonlinear Simultaneous Equations," Mathematics of Computation, 19, 577 (1965).

15. Friday, J. R. and B. D. Smith, "An Analysis of the Equilibrium Stage Separations Problem - Formulation and Convergence," AIChE J., 10, 689 (1964).

16. Bruno, J. A., J. L. Yanosik, and J. W. Tierney, "Distillation Calculations with Nonideal Mixtures," Extractive and Azeotropic Distillation, Advances in Chemistry Series 115, American Chemical Society, Washington, D.C., (1972).

17. Caillaud, J. B. and L. Padmanabhan, "An Improved Semi-Implicit Runge-Kutta Method for Stiff Systems," The Chem. Eng. Journal, 2, 227 (1971).

18. Feng, An, Ph.D. Dissertation, Texas A&M University, (1984).

19. Doukas, N. and W. L. Luyben, "Control of Sidestream Columns Separating Terney Mixtures," Instrum. Tech., 25, No. 6, 43 (1978).

20. Gallun, S. E. and C. D. Holland, "A Modification of Broyden's Method for the Solution of Sparse Systems - With Application to Distillation Problems Described by Nonideal Thermodynamic Functions," Comp. Chem. Eng. 4, 93 (1980).

21. Gallun, S. E. and C. D. Holland, "Solve More Distillation Problems, Part 5 - For Highly Nonideal Solutions," Hydrocarbon Processing, 55, No. 1, 137 (1976).

22. Gallun, S. E. and C. D. Holland, "Gear's Procedure for the Simultaneous Solution of Differential and Algebraic Equations with Application to Unsteady State Distillation Problems," Computers and Chemical Engineering, 6, No. 3, 231 (1982).

23. Gear, C. W., "Simultaneous Numerical Solution of Differential - Algebraic Equations," IEEE Transactions on Circuit Theory, 18, No. 1, 89 (1971).

24. Gear, C. W., Numerical Initial Value Problems in Ordinary Differential Equations, Prentice-Hall, Inc., Englewood Cliffs, N.J. (1971).

25. Greenstadt, J., Y. Bard and B. Morse, "Multicomponent Distillation Calculation on the IBM 704," Ind. Eng. Chem. 50, 1644 (1958).

26. Haas, J. R., C. D. Holland, F. Domingues and A. Gomez M., "Solution of Systems of Columns with Energy Exchange Between Recycle Streams," Comp. and Chem. Engr. 5, 41 (1980).

27. Hess, F. E., C. D. Holland, Ron McDaniel and N. J. Tetlow, "Solve More Distillation Problems, Part 7 - Absorber-Type Pipestills," Hydrocarbon Processing, 56, No. 5, 181 (1977).

28. Hess, F. E., S. E. Gallun, G. W. Bentzen, Ron McDaniel and N. J. Tetlow, "Solve More Distillation Problems, Part 8 - Which Method to Use," Hydrocarbon Processing, 56, No. 6, 241 (1977).

29. Holland, C. D. and A. Liapis, Computer Methods for Solving Separation Problems, McGraw-Hill Company, New York, (1983).

30. Haas, J. R., A. Gomez M. and C. D. Holland, "Generalization of the Theta Method of Convergence for Solving Distillation and Absorber-Type Problems," Sec. Sci. and Tech., 16, No. 1, 1 (1981).

31. Holland, C. D., Fundamentals of Multicomponent Distillation, McGraw-Hill Company, Inc., New York, (1981).

32. Holland, C. D., S. E. Gallun and M. J. Lockett, "Modeling Azeotropic and Extractive Distillations," Chem. Eng., 185 (March 23, 1981).

33. Holland, C. D., Fundamentals and Modeling of Separation Process, Prentice-Hall, Inc., Engelewoods Cliffs, N.J., (1975)

34. Holland, C. D., G. P. Pendon and S. E. Gallun, "Solving More Distillation Problems, Part 3 - Application to Absorbers," Hydrocarbon Processing, 54, No. 1, 101 (1975).

35. Holland, C. D., Unsteady State Processes with Applications in Multicomponent Distillation, Prentice-Hall, Inc. Englewood Cliffs, N.J., (1966).

36. Holland, C. D., Multicomponent Distillation, Prentice-Hall, Inc., Englewoods Cliffs, N.J., (1963).

37. Hooke, R and T. A. Jeeves, "Direct Search Solution of Numerical and Statistical Problems," J. Assoc. Comp. Mach., 8 No. 2 (1961)

38. Hutchinson, H. P. and C. F. Schewchuk, "A Computational Method for Multiple Distillation Towers," Trans. Inst. Chem. Eng., 52, 325 (1974).

39. Ischii, Y. and F. D. Otto, "A General Algorithm for Multistage Multicomponent Separation Calculations," Can. J. Chem. Eng., 51, 601 (1973).

40. Izarraraz, A., G. W. Bentzen, R. G. Anthony and C. D. Holland, "Solve More Distillation Problems, Part 9 - When Chemical Reactions Occur," Hydrocarbon Process, 59, 195 (1980).

41. Jelinek, J. and V. Hlavacek, "Steady State Countercurrent Equilibrium Stage Separation and Chemical Reaction by Relaxiation Method," Chem. Eng. Sci., 2, 79 (1976).

42. Kubicek, M., V. Hlavacek and F. Prochaska, "Global Modular Newton-Raphson Technique for Sumulation of an Interconnected System Applied to Complex Rectifying Columns," Chem. Eng. Sci., 31, 227 (1976).

43. Lewis, W. K., "Theory of Fractional Distillation," J. Ind. Chem., 1, 522 (1909).

44. Lewis, W. K. and G. L. Matheson, "Studies in Distillation - Design of Rectifying Columns for Natural and Refinery Gasoline," Ind. Eng. Chem., 24, 494 (1932).

45. Lord Rayleigh (J. Strutt), "On the Distillation of Binary Mixtures," Phil., Mag. 4, 527 (1902).

46. Lyster, W. N., S. L. Sullivan, D. S. Billinfsley, Jr. and C. D. Holland, "Figure Distillation This New Way Part 1 - New Convergence Method Will Handle Many Cases," Pet. Refinery, 38, No. 6, 221 (1959).

47. Mah, R. S. H., M. Michaelson, and R. W. H. Sargent, "Dynamic Behavior of Multicomponent Multistage Systems, Numerical Methods for Solution," Chem. Eng. Sci., 17, 6, 19 (1962).

48. McCabe, W. L. and E. W. Thiele, "Graphical Design of Fractionating Columns," Ind. Eng. Chem., 17, 605 (1925).

49. Michelsen, M. L., "An Efficient General Purpose Method of Integration of Stiff Oridnary Differential Equations," AIChE J., 22, No. 3, 594 (1976).

50. Michelsen, M. L., "Application of the Semi-Implicit Runge-Kutta Methods for Intergration to Ordinary and Partial Differential Equations," The Chem. Eng. Journal, 14, 197 (1977).

51. Mommessin, P. E., G. W. Bentzen and C. D. Holland, "Solve More Distillation Problems, Part 10 - Another Way to Handle Reactions," Hydrocarbon Process, 59, 144 (1980).

52. Murdoch, P. G. and C. D. Holland, "Multicomponent Distillation: IV - Determination of Minimum Reflux," Chem. Eng. Progr., 48, No. 6, 287 (1952).

53. Naphtali, L. M. and D. P. Sandholm, "Multicomponent Calculations by Linearization," AIChE J., 17, 148 (1971).

54. Newman, J. S., "Temperature Computed for Distillation," Petroleum Refiner, 42, 141 (1963).

55. Nelson, P. A., "Contercurrent Equilibrium Stage Separation with Reaction," AIChE J., 17, No. 5, 1043 (1971).

56. Peiser, A. M., "Better Computer Solution of Multicomponent Systems," Chemical Engineering, 67, 1929 (July 1960).

57. Ponchon, Marcel, "Graphical Study of Distillation," Tech. Modern, 13, 20 (1921).

58. Prokopakis, G. J. and W. D. Seider, "Dynamic Simulation of Distillation Towers," AIChE Annual Meeting, Chicago, Illinois, (November 1980).

59. Ricker, N. L. and E. A. Grens, "A Calculational Procedure for Design Problems in Multicomponent Distillation," AIChE J., 20, No. 2, 238 (1974).

60. Robinson, C. S., Elements of Distiliation, listed, McGraw-Hill Book Co., Inc., New York, (1962).

61. Robinson, C. S. and E. R. Gilliland, Elements of Fractional Distillation, 4th Ed., McGraw-Hill Book Company Inc., New York, (1950).

62. Roche, E. C. and H. K. Staffin, "Rigorous Solution of Multicomponent Multistage Liquid-Liquid Extraction Problems," paper presented at 61st Annual AIChE Meeting, Los Angeles, California, (December 1969).

63. Rose, A., R. F. Sweeny and V. N. Schrodt, "Continuous Distillation Calculation by Relaxation Method," Ind. Eng. Chem., 50 737 (1958).

64. Rose, B. A. and W. D. Seider, "Simulation of Three-Phase Distillation Towers," Computers and Chem. Eng., 5, No. 1, 7 (1981).

65. Sargent, R. W. H. and K. Gaminibandara, "Optimum Design of Plate Distillation Columns," Optimization in Action, p. 266, L. C. W. Dixon ed., Academic Press, New York, (1976).

66. Schubert, L. K., "Modification of Quadi-Newton Method for Nonlinear Equation with a Sparse Jacoblian," Math. Comp., 25, 27 (1970).

67. Suzuki, I., H. Komatsu and M. Hirata, "Formulation and Prediction of Quaternary Vapor-Liquid Equilibria Accompanied by Chemical Reaction," Journal of Chem. Eng. of Japan, 3, No. 2, 152 (1970).

68. Sorel, E., La Rictification de l' Alcoal, Ganthier-Villais et fils, Paris, France, (1893).

69. Sujata, A. D., "Absorber Stripper Calculations Made Easier," Hydrocarbon Processing, 40, 137 (1961).

70. Srygley, J. M. and C. D. Holland, "Optimum Design of Conventional and Complex Columns," AIChE J., 11, No. 4, 695 (1965).

7. Thiele, E. W. and R. L. Geddes, "Computation of Distillation Apparatus for Hydrocarbon Mixtures," Ind. Eng. Chem., 25, 289 (1933).

72. Tierney, J. W. and J. L. Yanosik, "Simultaneous Flow and Temperature Correction in the Equilibrium Stage Problem," AIChE J., 15, 897 (1969).

73. Tierney, J. W. and J. L. Yamosik, "Simultaneously Flow and Temperature Correction in the Equilibrium Stage Problem," AIChE J., 6, 897 (1969).

74. Tierney, J. W. and J. A. Bruno, "Equilibrium Stage Calculations," AIChE J., 13, 556 (1967).

75. Tomme, W. J. and C. D. Holland, "Figure Distillation This New Way: Part II - When Columns are Operated as a Unit, Petroleum Refiner, 41, No. 6, 139 (1962).

76. Tomich, J. F., "A New Simulation Method for Equilibrium Stage Processes", AIChE J., 16, 29 (1970).

77. Tung, L. S., T. F. ldgar, "Development and Reduction of a Multivariable Binary Distillation Column with Tray Hydraulics", AIChE National Meeting, Houston, Texas, (April 1979).

78. Underwood, A. J. V., "Fractional Distillation of Multicomponent Mixtures," Chem. Eng. Progr., 44, 603 (1948).

79. Waggoner, R. C. and C. D. Holland, "Problems Involving Conventional and Complex Columns at Unsteady State Operation," AIChE J., 11, 112 (1965).

80. Young, S., Fractional Distillation, 1st ed., The McMillan Company, New York, (1923).

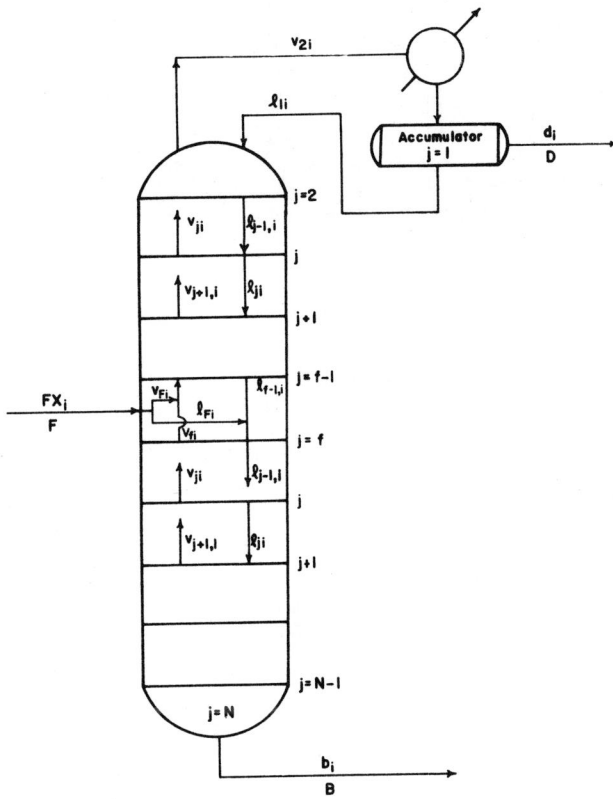

Figure 1. Conventional distillation column and identifying symbols.

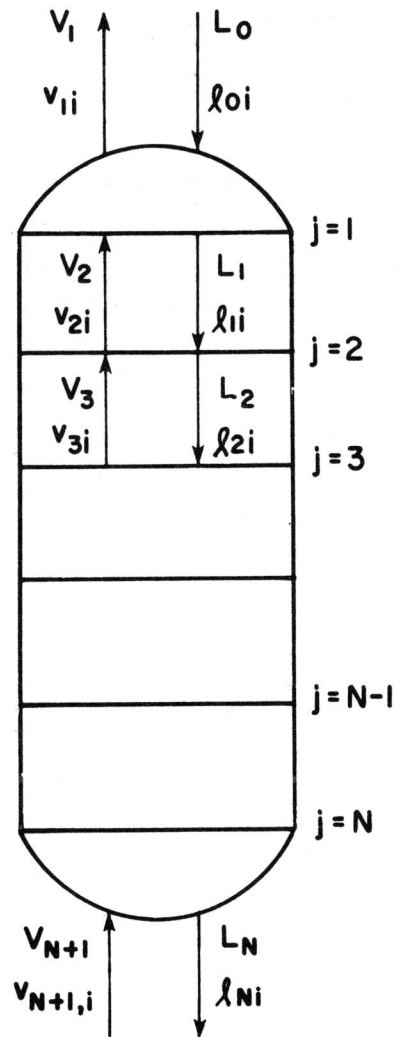

Figure 2. Absorber and identifying symbols.

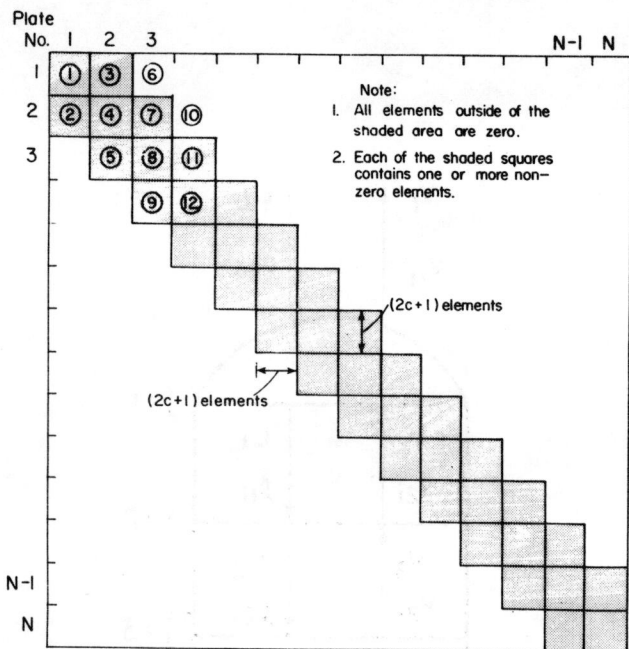

Figure 3. Structure of the Jacobian matrix for an absorber.

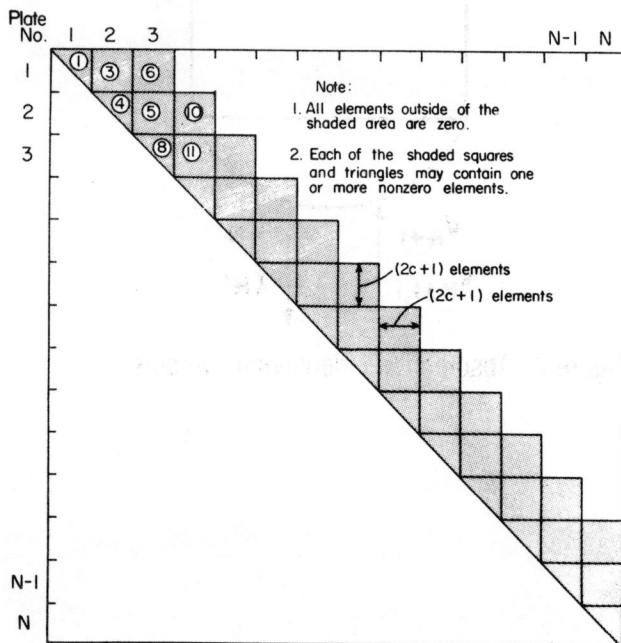

Figure 5. A system of two columns with heat exchange between recycle streams.

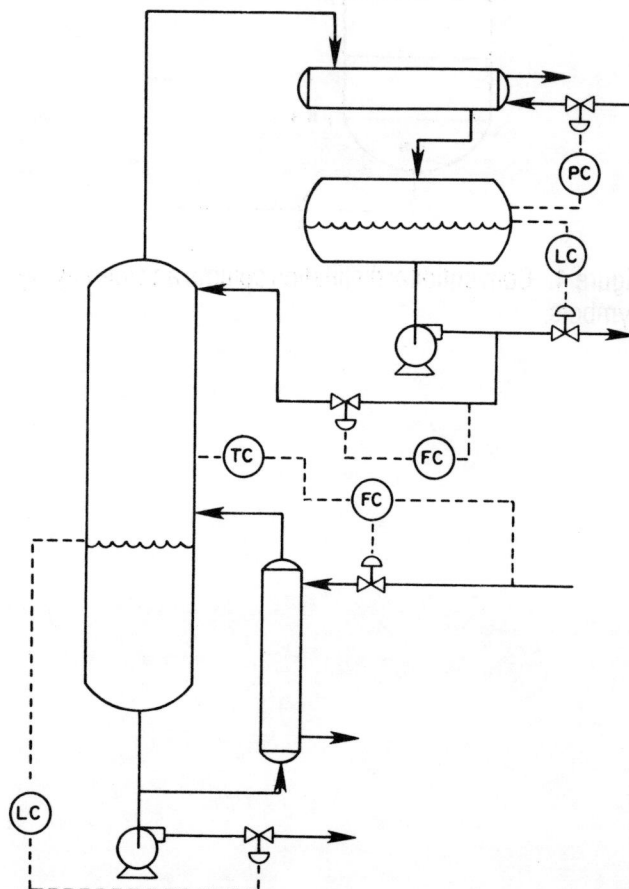

Figure 4. Upper triangular matrix for an absorber.

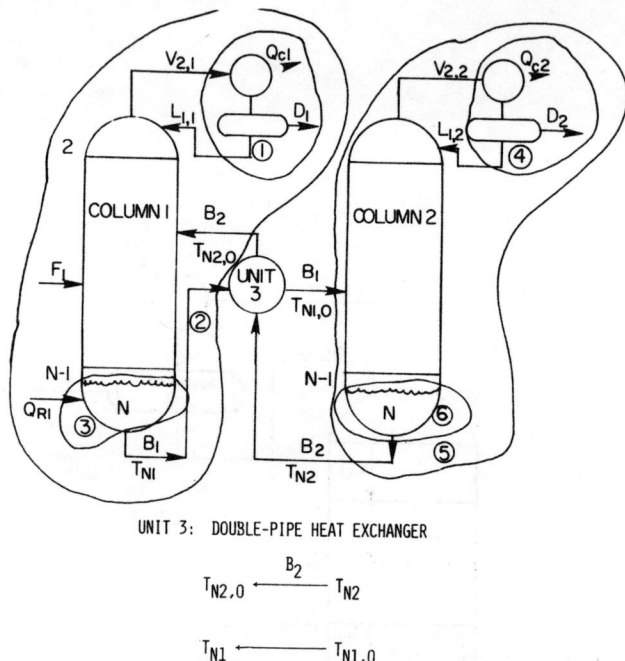

Figure 6. Tower control system.

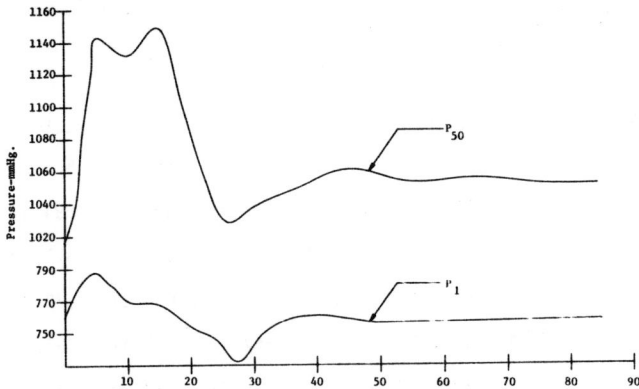

Figure 7. Response of receiver and base pressure.
(p_1 = pressure in accumulator and p_{50} = pressure in reboiler.)

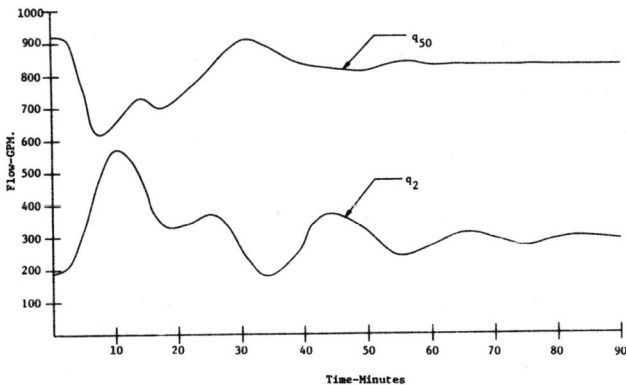

Figure 8. Response of bottom and distillate total flow
(Q_{50} = bottoms rate; Q_2 = distillate rate)

Table 1. State of Example 1.

Component	FX_i
CH_4	2.0
C_2H_6	10.0
C_3H_6	6.0
C_3H_8	12.5
$i-C_4H_{10}$	3.5
$n-C_4H_{10}$	15.0
$n-C_5H_{12}$	15.2
$n-C_6H_{14}$	11.3
$n-C_7H_{16}$	9.0
$n-C_8H_{18}$	8.5
400*	7.0

Specifications

$D = 31.5$, $V_2 = 94.8$ (all flow rates are in lb mole/hr
partial condenser, column pressure = 300 lb/in^2 abs,
$N = 13$ and $f = 5$. Equilibrium and enthalpy data for all
components are given in Tables B-1 and B-2 of Reference (31).
The initial temperature profile is to be taken linear
with plate number between $T_1 = 610°R$ and $T_{13} = 910°R$. Take
the initial vapor rate profiles to be $V_j = 94.8$ ($j = 2$,
3, ..., 13) and the corresponding liquid rate profile is
given by material balance. Component $i-C_4H_{10}$ was taken
as the base component and a and b in Equation (14) were
determined on the basis of the values for the K of
$i-C_4H_{10}$ at 510 and 960°R.

*Commonly referred to a the 400°F - normal boiling fraction.

Table 2. Temperature profiles, theta, and calculated values for Example 1.

Temperature profiles (°R)
Trial

STAGE	1	2	3	4	5	6	7	8	9	10	11	12
1 (distillate)	573.59	567.94	567.56	567.56	567.56	567.57	567.57	567.57	567.57	567.57	567.56	567.57
2	596.60	590.67	594.54	594.13	594.42	594.33	594.38	594.36	594.37	594.37	594.37	594.37
3	607.01	606.81	612.61	611.63	611.99	611.89	611.95	611.92	611.93	611.93	611.93	611.93
4	612.65	627.85	631.10	629.91	630.26	630.25	630.27	630.26	630.27	630.27	630.27	630.26
5 (feed)	617.02	695.72	658.24	670.53	665.95	668.04	667.14	667.53	667.37	667.44	667.41	667.41
6	647.02	709.68	681.56	691.68	687.53	689.20	688.47	688.79	688.65	688.71	688.68	688.69
7	670.05	717.92	697.73	705.76	702.37	703.69	703.11	703.36	703.25	703.30	703.28	703.28
8	689.59	724.20	709.70	716.11	713.40	714.46	714.00	714.19	714.11	714.14	714.13	714.13
9	707.47	730.07	719.35	724.47	722.32	723.17	722.79	722.95	722.88	722.91	722.90	722.90
10	725.93	737.58	728.94	732.96	731.25	731.92	713.63	731.75	731.70	731.72	731.71	731.71
11	747.73	749.93	742.09	745.17	743.78	744.32	744.08	744.18	744.13	744.15	744.15	744.15
12	777.47	774.19	766.54	768.84	767.70	768.13	767.94	768.02	767.99	768.00	767.99	768.00
13 (bottoms)	833.68	833.41	825.47	827.21	826.36	826.68	826.53	826.59	826.56	826.58	826.57	826.57
D (calculated)	31.6175	31.5925	31.6031	31.5986	31.60057	31.59977	1.01356	0.994182	1.00243	0.998940	1.00042	1.00000

Table 3. Vapor rates obtained by the constant composition method for Example 1.

Vapor rates (lb-mole/hr)
Trial

STAGE	1	2	3	4	5	6	7	8	9	10	11	12
2	94.80	94.80	94.80	94.80	94.80	94.80	94.80	94.80	94.80	94.80	94.80	94.80
3	94.63	93.23	93.06	93.29	93.29	93.29	93.29	93.29	93.29	93.29	93.29	93.29
4	94.13	88.11	89.31	89.41	89.45	89.41	89.41	89.41	89.42	89.42	89.42	89.42
5	93.13	72.10	82.71	79.25	80.51	79.94	80.18	80.08	80.12	80.10	80.11	80.11
6	140.71	108.19	117.18	109.35	111.52	110.55	110.92	110.75	110.82	110.79	110.81	110.80
7	143.80	125.43	129.81	125.40	126.79	126.13	126.36	126.25	126.29	126.24	126.28	126.27
8	146.69	137.24	138.81	136.34	137.20	136.78	136.91	136.84	136.87	136.86	136.86	136.86
9	148.52	145.60	145.39	144.16	144.60	144.38	144.44	144.41	144.42	144.42	144.42	144.41
10	148.03	150.07	149.13	148.83	148.94	148.87	148.89	148.87	148.88	148.88	148.88	148.87
11	144.63	150.08	149.09	149.59	149.43	149.50	149.47	149.48	149.47	149.48	149.47	149.47
12	137.74	143.88	142.75	143.97	143.60	143.77	143.70	143.73	143.72	143.72	143.72	143.72
13	120.86	123.86	122.24	123.93	123.33	123.60	123.49	123.54	123.52	123.53	123.52	123.52

Table 4. Solution sets of the component flow rates in the distillate and bottoms for Example 1.

Component	d_i	b_i
CH_4	0.200000×10	0.11163×10^{-8}
C_2H_6	0.999990×10	0.11627×10^{-3}
C_3H_6	0.597230×10	0.27665×10^{-1}
C_4H_8	0.1234600×10^2	0.15358
$i-C_4H_{10}$	0.74216	0.27578×10
$n-C_4H_{10}$	0.53699	0.14462×10^2
$n-C_6H_{12}$	0.94035×10^{-5}	0.11299×10^2
$n-C_7H_{14}$	0.94025×10^{-5}	0.89999×10
$n-C_8H_{16}$	0.63427×10^{-7}	0.84999×10
400	0.65162×10^{-2}	0.69999×10

$Q_C = 3.9628 \times 10^5$ Btu/hr

$Q_R = 1.3278 \times 10^6$ Btu/hr

Convergence criterion: $|g(1)| \geq 10^{-5}$

Table 5. Statement of Example 2.

Component	F_1X_{1i} (g mole/sec)	F_2X_{2i} (g mole/sec)	F_3X_{3i} (g mole/sec)
Methyl alcohol	0	0.25	65.0
Acetone	0	0.50	25.0
Ethanol	0	5.0	5.0
Water	5.0	197.5	5.0

Other Specifications

The column has a total condenser, 50 stages, F_1 enters on plate 4, F_2 on plate 6, and F_3 on plate 21. The column pressure is 760 mm. Feeds F_1, F_2, and F_3 enter with enthalpies 0.33993877×10^4, 0.2853018×10^6, and 0.08139422 cal/g mole, respectively. A reflux ratio (L_1/D) of 2.5 is to be used, and the bottoms is to be withdrawn at the rate of 285 lb mole per hour. The equilibrium and thermodynamic data to be used are the same as stated for Example 5-3 of Reference (31).

Table 6. Solution of Example 2.

1. Final profiles, temperature, and vapor and liquid rates

PLATE	T_j (°F)	V_j (lb mole/hr)	L_j (lb mole/h)	PLATE	T_j (°F)	V_j (lb mole/hr)	L_j (lb mole/hr)
1	134.36	·····	58.13	26	154.61	78.21	36.32
2	136.11	81.38	56.32	27	154.61	78.21	36.32
3	139.12	79.57	53.77	28	154.61	78.21	36.32
4	144.65	77.02	56.31	29	154.61	78.21	36.32
5	152.70	74.56	52.30	30	154.61	78.21	36.32
6	169.06	70.55	25.66	31	154.61	78.21	36.32
7	169.33	71.58	25.66	32	154.61	78.21	36.32
8	169.50	71.60	25.66	33	154.61	78.21	35.62
9	169.62	71.61	25.66	34	154.61	78.21	36.32
10	169.72	71.61	25.66	35	153.62	78.21	36.32
11	169.81	71.61	25.66	36	154.62	78.21	36.32
12	169.92	71.61	25.66	37	154.62	78.21	36.32
13	170.05	71.61	25.66	38	154.62	78.21	36.32
14	170.23	71.60	25.66	39	154.62	78.21	36.32

Table 6. Continued.

15	170.45	71.60	25.66	40	154.63	78.21	36.32
16	170.70	71.60	25.66	41	154.65	78.20	36.32
17	170.91	71.61	25.67	42	154.69	78.19	36.32
18	170.84	71.65	25.68	43	154.77	78.17	36.31
19	169.96	71.78	25.68	44	154.91	78.14	36.31
20	166.64	72.15	25.72	45	155.20	78.07	36.29
21	154.61	73.55	25.86	46	155.75	77.94	36.27
22	154.61	78.21	36.32	47	156.77	77.71	36.23
23	154.61	78.21	36.32	48	158.63	77.32	36.17
24	154.61	78.21	36.32	49	161.93	76.73	36.06
25	154.61	78.21	36.32	50	169.48	75.64	28.50

2. Product Distribution

Component	b_i (lb mole/hr)	d_i lb mole/hr)
Methyl alcohol	0.8832228×10^{-1}	0.6516167×10^{2}
Acetone	0.1936700×10^{2}	0.6133089×10
Ethanol	0.1895224×10	0.8104776×10
Water	0.1899549×10	0.2056004×10

$Q_C = 111,504.0$ Btu/hr

$Q_R = 126,209.9$ Btu/hr

3. Comparison of the Broyden-Householder and the Schubert Algorithms

Method	Method of caluclation of derivatives	Evaluation	Factorization	Iterations	Final squared norm	Execution time (sec)
Newton-Raphson	Analytical	14	13	13	7.2×10^{-13}	9.97
Newton-Raphson	Numerical	13	12	12	4.0×10^{-11}	59.94
Broyden-Householder	Analytical	4	4	51	197.0×10^{-10}	12.77
Broyden-Householder	Numerical	5	5	56	1.16×10^{-9}	35.90
Schubert	Analytical	1	37	37	8.91×10^{-10}	22.77
Schubert	Numerical	1	34	34	2.40×10^{-10}	26.89

*AMDAHL, FORTRAN H, OPT 2

Table 7.

1. Single Columns							
Example				Execution times and number of trials on an AMDAHL 470V/6 computer			
				2N Newton-Raphson			
Type of Column	No. of Stages	No. of Components	θ Method	Procedure 1	Procedure 2	Procedure 3	Compiler
Conventional distillation column	13	11	2.6 s 12 trials*	10.74 s 5 trials*	12.45 10 trials* 2 = no. of evaluations of jacobian matrix**	12.98 s 12 trials 2	WATFIV
Complex distillation column	52	7	2.14 19	39.18 11	27.0 11 3	23.42 11 3	FORTRAN 3 with OPT 2
Complex distillation column	13	11	2.81 13	13.10 6	12.57 9 2	11.93 9 2	WATFIV
Absorber	8	14	....	6.22 5	6.25 18	8.45 15	WATFIV
Reboiler absorber	11	6	....	9.91 9	7.46 11 2	7.13 12 2	WATFIV
Stripper	8	9	....	7.09 9	7.29 14	7.16 14	WATFIV

* For the θ method, the following norm ϕ was used as the convergence criterion, $\phi = 1/n \sum_{i=1}^{n} |g_i(1,1,\ldots,1)|$, where n is equal to the number of g functions. For the 2N Newton-Raphson method, $\phi = 1/N [\sum_{i=1}^{N} f_j^2]^{1/2}$, where N is equal to the number of functions f. For all examples except ϕ was taken to equal to 10^{-4} or 10^{-5}.

Table 7. Continued.

2. Systems of Columns				
Type of Column	No. of Components	No. of Stages	Method	Compiler
System of distillation column	7	104	1. The o and θ methods: 3.65 s, 17 system trials. 2. Procedure 3 of the 2N Newton-Raphson method: 90 s per iteration did not run to convergence.	FORTRAN H with OPT. 2
Pipestill	35	37	Procedures 1, 2, and 3 of 2N Newton-Raphson method. Procedure 1:214.59 s, 15 trials, Procedure 2: 145.23 s, 13 trials. Procedure 3: 140.23 s, 12 trials, jacobian matrix evaluated 3 times for procedures 2 and 3.	FORTRAN H with OPT. 2
Reboiled absorber and distillation	12	24	2N Newton-Raphson method Procedure 3 for reboiled absorber, θ method for distillation column and Θ method for the system. Procedure 1: 17.8s, 15 trials. Procedure 2: 14.62 s, 18 trials Procedure 3: 14.46 s, .8 trials.	FORTRAN H

** The number of evaluations of the jacobian matrix includes the initial evaluation cailed for in step 2 plus any additional evaluations required in step 4 of the calculational procedure described for Broyden's method.

Table 8. Recommended calculational procedures.

Type of Column	Type of Fluid	Recommended Procedure
Distillation, conventional or complex with both condenser and reboiler	Ideal or near ideal solutions	θ method
Absorbers, strippers, reboiled absorbers or any column without a condenser or a reboiler	Ideal solutions	2N Newton-Raphson method
Any type of column	Highly nonideal	Almost band algorithm
Systems of columns, inter-connected by mass and/or energy recycle streams	Use procedure recommended above for the individual columns; use the capital θ method for the system.	

Table 9. Elements of the vector L and values of β_{-1}

Elements of L	Order: k - 1, 2, . . . , 6					
	1	2	3	4	5	6
0	1	$\frac{2}{3}$	$\frac{6}{11}$	$\frac{24}{50}$	$\frac{120}{74}$	$\frac{720}{1764}$
1	1	$\frac{3}{3}$	$\frac{11}{11}$	$\frac{50}{50}$	$\frac{274}{274}$	$\frac{1764}{1764}$
2		$\frac{1}{3}$	$\frac{6}{11}$	$\frac{35}{50}$	$\frac{225}{274}$	$\frac{1624}{1764}$
3			$\frac{1}{11}$	$\frac{10}{50}$	$\frac{85}{274}$	$\frac{735}{1764}$
4				$\frac{1}{50}$	$\frac{15}{274}$	$\frac{175}{1764}$
5					$\frac{1}{274}$	$\frac{21}{1764}$
6						$\frac{1}{1764}$
β_{-1}	1	$\frac{2}{3}$	$\frac{6}{11}$	$\frac{12}{25}$	$\frac{60}{137}$	$\frac{60}{147}$

COMPUTERS IN CHEMICAL ENGINEERING ADVANCES AND ACCOMPLISHMENTS

Stanley I. Proctor ■ Monsanto Company, St. Louis, MO

Computers have played a significant role in process engineering in recent years. In the future, as hardware and software becomes more reliable, less expensive, and more versatile, this role will expand. This presentation will review the developments in computer hardware and software which have had an impact on advances in chemical process engineering. The most significant accomplishments in chemical engineering computing and systems technology which have resulted will be discussed. Finally, the impact of these developments on the profession, the chemical industry, and society in general will be considered.

DEVELOPMENTS IN COMPUTER HARDWARE AND SOFTWARE

In the last seventy-five years, the chemical industry has come a long way. Indeed, since the Second World War, the growth has been phenomenal. The same can be said of the computer industry. Although one can trace the roots of the computer industry back over a hundred years, the most significant advances have been in the last twenty-five or so.

The beginning of modern computing technology is usually considered to start with Charles Babbage, an English mathematician, who spent his life trying to develop a device to carry out mathematical calculations. In about 1834 he proposed constructing a steam-driven Analytical Engine possessing many of the characteristics of a modern day digital computer.

He planned to use punched cards to control his device. He borrowed this idea from Joseph Jacquard, who, in 1801, automated a loom using punched cards to control the loom and define the pattern in the cloth.

Babbage never quite completed any of his engines. Some say that he was ahead of existing technology. But one of his major problems was himself. He could never learn to freeze his design and stop trying to make changes.

I would be remiss if, in talking about Babbage, I did not mention Countess Ada Augusta Lovelace, the daughter of the famous poet, Lord Byron. She was a mathematician in her own right, worked closely with Babbage, and planned his computational problems. She has been called the world's first programmer.

Sometime later, Herman Hollerith invented the tabulating machine which allowed the U.S. Census Bureau to automate the 1890 census. It also used punched cards. Even with a substantial growth in population, the tabulations were finished in three years, whereas the previous census had taken nine.

The modern computer is largely an outgrowth of military requirements in World War II. The Army, especially, worked on trying to improve the differential analyzers developed by Vannevar Bush and used at Aberdeen Proving Grounds to calculate ballistic tables. It took two to three months to complete a table.

John Mauchley and Presper Eckert of the Moore School of Engineering at the University of Pennsylvania felt a digital computer could do better. They submitted a proposal in 1943 for a device that would compute a trajectory and complete a table in two days. The ENIAC (Electronic Numerical Integrator and Calculator) was constructed at the expenditure of 200,000 manhours. It contained no moving parts except input/output gears. It had 500,000 soldered joints, 18,000 vacuum tubes, 6,000 switches, and 500 terminals. It is said that the lights in Philadelphia dimmed when all 18,000 of these vacuum tubes came on.

At about the same time, other efforts were underway. Howard Aiken and co-workers developed the MARK I which went into operation at Harvard in 1944. John von Neumann, in 1945, developed the design for the EDVAC Electronic Discrete Variable Automatic Computer) at Princeton, incorporating the concept of a "stored program".

Over the next fifteen or so years computers continued to evolve. Much of the application was business related although some technical and scientific calculations were carried out. This brings us to the last half of the decade of the 50's and, about this time, the significant application of computers to chemical engineering calculations began.

Since that time, what have been the developments which have most significantly impacted on chemical engineering?

In software, the most significant has been the development of higher level computer languages. In 1957, John Backus of IBM and his team developed the FORTRAN language. Prior to that, programs had to be written in assembly language which meant that one instruction had to be written for each machine operation.

FORTRAN has made computers much more accesible to engineers and other non-computing professionals. It is probably still the most widely used of all the scientific programming languages. It has enabled the engineer to communicate with computers using a language which is a hybrid of mathematical expressions and English statements. Each FORTRAN expression is then converted (compiled) into a number of machine instructions. This reduction in the number of required expressions results in improved productivity and reduced programming time.

A further increase in productivity was the development of problem oriented languages. This development began in the 60's and is continuing today. These languages allow the engineer to specify his problem or program definition in terminology closely akin to common engineering usage.

Assembly language required a computer professional to program the computer. FORTRAN gave programming access to the sophisticated engineering user. The problem oriented languages now give the casual engineering user the ability to structure a computer program to solve his specific problem.

Flowing from the development of languages have been advances in applied mathematics and numerical analysis. The continuing improvement of methods for equation solving and optimization, making them more reliable and robust, faster and more flexible, has enabled engineers to solve more complex problems with a greater degree of rigor in less time than ever before.

Contributing to this have been the significant developments in computer hardware. The microelectronics revolution has had far reaching implications. We have gone from vacuum tubes to transistors to chips where thousands of bits of information or instructions can be placed on a device no larger than a paper clip, with access times on the order of nanoseconds. Low cost mass storage devices such as discs have replaced cards and paper tapes. Magnetic tapes, although still heavily used for archival storage, have also been replaced for on-line storage.

As a result, computer speed has been increasing exponentially, physical size requirements for memory have been decreasing and the costs of computing has steadily declined.

All of these factors have lead computer hardware in two directions. The first is toward larger and faster machines with hugh amounts of main and auxiliary storage. These monster "number crunchers" are capable of handling massive amounts of data and performing the most complex of calculations.

The second is toward smaller, more efficient machines. We have seen the development of the minicomputer, the microcomputer, and now the personal computer. Each has had and will have its impact on chemical engineering. Without a doubt, the personal computer is currently the hottest item on the market. When

Time magazine makes a computer its "Man of the Year" you know that it is really "in". It has been estimated that there are four million personal computers in homes and about two and one-half million in businesses today. The forecast is for eighty million to be in use by the end of this century. That has got to have some impact on the profession, the individual engineer, and the educational system in which that engineer is trained. But, just what that will be is not entirely clear at this time. There is a lot of media hype concerning personal computers and the dust needs to settle a bit before we can see our way clearly.

Another development related to personal computers and brought about by the microelectronics revolution is in the computer power of hand held or portable calculators. Portable calculators are not new. In 1888, William Seward Burroughs received his first patent for a calculator and in the next year he had manufactured fifty machines. Unfortunately, they proved impossible for anyone but Burroughs to operate and they were recalled. There was one exception, a field agent who operated his calculator so well that he refused to return it, preferring to haul it from saloon to saloon in a wheelbarrow betting drinks on his mathematical accuracy. It wasn't hand held, but it was portable.

Not only has portability been improved since then, but so have other features. In less than a single generation, the slide rule has all but disappeared, giving way to increased accuracy of the electronic calculator with its correct placement of the decimal. Remember the days when you had to mentally compute the decimal point! Costs have significantly declined and capabilities increased to the point where the application of the "computer on a chip" has given some programmable calculators today the computing power equivalent to large scale digital computers of twenty-five to thirty years ago.

Another item which I would like to mention is the advent of interactive computing coupled with networking. Interactive computing has enabled the user to interface directly, through a terminal, with the computer, and has permitted the transfer of data and information among computers or between the computer and external devices (as in process control, for example).

Finally, the development of a computerized graphics capability brought about by hardware advances in CRT terminals and software advances in data handling has resulted in a powerful tool

which is only beginning to be utilized. The potential applications in both the scientific and business areas of our industry are enormous.

ADVANCES AND ACCOMPLISHMENTS IN CHEMICAL ENGINEERING

As we have seen, there has been a phenomenal evolution in computer hardware and software in the past twenty-five years. What has been the impact of that on chemical engineering? Can we relate any significant advances in chemical engineering to these computer developments? The answer is, of course, yes, or this paper would not be necessary.

When we consider where computers have had the greatest impact in chemical engineering, four areas come immediately to mind: process control, process design, information storage and retrieval, and graphics. To be sure, graphics is an integral part of the first three but since it is so pervasive, I will discuss it as a separate item, also.

An important application for digital computers and one which is increasing rapidly, is that of process control. After initial major strides in the late fifties and early sixties, growth and development appeared to have leveled out. From the mid-seventies, however, the pace quickened again.

Let us look at what sets a process control computer apart from other digital computers: First of all, it is sensor based. This means that it gets its input and performs its function by measuring the values of a set of selected variables. The analog realtime signals are converted to digital values on which computations and comparisons are performed. The results are the necessary corrections to the operating parameters that will cause the process to reach or maintain the desired operations level. In effect, the successful process control system turns the process into a "self-correcting machine".

Second, it is event-driven. This means that sequences are initiated and actions are carried out as a result of outside occurrences from the process at the time they occur, or by a clock built into the computer. This is in sharp contrast to the early computer systems, which were batch oriented.

The event-triggered or realtime operation necessitates much more sophisticated operating systems. The process control computer has to be able to, or at least appear to be able to, handle many tasks at the same time. In addition, it has to be able to react to interruptions from the process and determine the priority of the tasks that need to be processed at any one time.

The development of the minicomputer in the late fifties opened the opportunities for process computer control. The first installation of a control computer in a chemical plant was in 1960. It was a Ramo-Wooldridge RW-300 by Monsanto Company at its ammonia plant in Luling, Louisiana. Developments in control computers have paralleled the computer industry overall - faster, cheaper, more storage, more reliable. Control can be of different types - single or multi-loop, supervisory or direct digital control (DDC). Although it has been discussed for many years, the rapid advances in microelectronics has now brought us to the reality of hierarchical distributed control with the potential of plant-wide information and control networks.

In the early days of computer control, the primary effort was in simply trying to prove that it could be done. Over the years, the viability of chemical process computer control has been demonstrated to the degree that in many instances on new installations today the question is not "why install a computer?" but rather "why not?" Some of the demonstrated benefits include:

- Improved yields.
- Increased throughput.
- Shortened batch times.
- Lower labor cost.
- Higher purity.
- Lower energy use.
- Overall cost minimization.
- Improved safety.
- Decreased maintenance costs.

One of the most significant accomplishments in chemical engineering computing applied especially to process design has been the development of process simulation technology. This includes both steady-state and dynamic simulation. Both are important, but probably the most far reaching impact has been in steady-state simulation, or flow sheeting as it is sometimes called.

Flowsheet simulation enables the engineer to develop a flowsheet by selecting the units involved in a process and the interconnecting streams. By specifying the parameters which define the unit and choosing the operating conditions of the process, a complete mathematical model of the process is generated. The equa-

tions are then solved to yield a material and energy balance for the process. By manipulating unit parameters and operating conditions the engineer can develop his design and evaluate sensitivities. By changing the units or their location in the flowsheet, design alternatives can be evaluated.

The development of flowsheet simulators began in the late 1950's. The work has progressed to the point that, in 1960, the AIChE Machine Computation Committee, a forerunner of the CAST Division, held an all day workshop on computer calculation of heat and material balances. This development has been strongly influenced by developments in computer software, including both advances in computer languages and numerical methods for equation solving. The trend toward larger and faster hardware has permitted increased rigor and complexity in these simulators. This has broadened the area of applicability and improved the accuracy of design resulting from the use of these tools.

One of the earliest simulators was PACER, written by Shannon and Mosler at Purdue. At the same time, many chemical and petroleum companies were also developing flowsheet simulators for internal use. Most well known of these is FLOWTRAN, developed by Monsanto, because it has become available publicly. Many other simulators have been developed and some have also become commercially available in the seventies and eighties. In addition to FLOWTRAN, some better known ones are ASPEN, developed at MIT and now marketed as ASPEN Plus by ASPEN Tech; CONCEPT, from the Computer Aided Design Center in England; DESIGN/2000, from Chem Share; and PROCESS, from Simulation Sciences.

Most flowsheet simulators consist of several key elements. The Preprocessor converts the user "language" to a FORTRAN program which is compiled and linked with those routines needed to simulate individual blocks and to generate required physical property data.

One of the most significant features of a simulator is the physical property system. Since perfect accuracy in the representation of physical properties of any real compound is impossible, there are two things necessary in a simulation tool. First, it is necessary to use the best physical property correlations available so that reasonably good calculations can be made from incomplete data, and, second, provided that accurate data are available, then various correlations should be able to fit the data with close agreement.

In addition to the programs for data correlation and estimation, a simulator usually contains a Physical Property Data Base for storage and retrieval of needed physical property information.

A Unit Operations Library generally contains an extensive set of "building blocks" to describe a wide range of unit operations. The user has several choices among rigorous and shortcut techniques. In conjunction with the unit operation blocks there may be cost and sizing blocks that allow the user to determine preliminary cost and sizing information on selected units or complete processes.

Flowsheet simulators have revolutionized the way a chemical engineer carries out a process design. They have essentially automated the calculational aspects of the design function. The design process, and the engineering judgement required in the process, have not been eliminated, but the power of the computer has been put at the disposal of the process design engineer to significantly expand his/her capability.

Flowsheet simulators have gained such wide acceptance that I suspect that most process designs of consequence involving vapor-liquid processes use this tool. And with the increasing ability of simulators to handle solids, electrolytes, polymers, etc., the use has expanded, not only within the "classical" chemical industry, but in related industries as well. They have been used successfully in an enormous variety of applications, including:

- New Process Design
- Process Feasibility Studies
- Energy Conservation Studies
- Optimization of Capital or Operating Costs
- Evaluation of Alternate Feedstocks
- Pilot Plant Scaleup
- Debottlenecking Studies
- Efficiency Studies
- Optimization of the Location of Recycle Streams
- Evaluation of Competitive Processes
- Determine Process Variable Sensitivities
- Determination of the Pollution Effects of a New or Revised Process

Although the principal impact of simulation on the chemical industry has been through steady-state simulation, I should also mention dynamic simulation. Dynamic simulation, that is, the ability to mathematically represent the non-steady-state conditions of process operation, is very important in considering batch operations, or process control, or kinetics modeling.

Dynamic simulation of chemical processes began in the late fifties with the advent of analog computers. These devices, which were first used in the late forties and early fifties for military applications, used electrical circuitry, and the equations thereof, as an "analog" of the physical and chemical processes to be represented or simulated. In the late sixties these systems, which had evolved into hybrid computers (a combination of analog and digital computers), were superseded by totally digital systems. High speed hardware and software systems with efficient equation-solving methods coupled with high level language capability have enabled the development of software products which have become standard for dynamic simulation.

Dynamic simulation falls in between and overlaps and connects the two previously discussed areas of process control and process simulation. The primary advantage of this tool is to enable the engineer to efficiently and economically design reactor systems and process control loops. The dynamic simulation of large chemical plants is still largely not possible today, however.

The use of computers for information storage and retrieval has had broad application in chemical engineering. The increase in storage capability coupled with rapid and efficient access to items stored has, in conjunction with software developments in data base management, given the engineer ready access to information. Instead of laboriously digging through hard copy to obtain information, the engineer need only review computer files via key words to obtain references to sources of interest. Physical property data, until recently available only through books and compilations, is now available, via computer, not only in numerical but in graphical form. Physical property data bases are available, not only as part of in-house or commercial process simulators, but as independent systems.

This capability has placed at the disposal of engineers more information than ever before, and has, as a result, improved the efficiency, accuracy, and reliability of their work. When information storage and retrieval is combined with computer networking, the impact on the communications capability among engineers is tremendous. They can now easily share information or pass information to each other rapidly and thus improve their productivity and effectiveness.

Finally, I have mentioned graphics in terms of physical property information. The use of graphics is pervasive throughout engineering. Graphics is the natural language of the engineer, be it drawings, diagrams, plots, or pictures. A picture is indeed worth a thousand words. The advances in on-line computation, mass storage capability and lower cost in computation have placed on-line graphics capability in the hand of the engineer. Whether it is used for management and control information, for display of technical data, or for generation of flowsheets, as a tool for engineering applications it is only beginning to be exploited. For example, in computer aided engineering, graphics is becoming a key vehicle for information display and transfer. This tool promises to be an extremely effective device. The visual impact of information display is significant and should not be under estimated.

THE IMPACT OF COMPUTERS IN CHEMICAL ENGINEERING

What has been the impact of all of these computer developments? on the chemical engineer? on the chemical engineering profession? on the chemical industry? on society as a whole?

The major impact of computers to date has not been to influence to a large extent what engineers do - the design process, for example, is similar to what it was twenty-five years ago - but it has influenced to a great extent how they do it. They don't spend days on calculations with slide rules and handle volumes of calculation sheets as design support information. They don't need to apply large safety factors in their designs and more often than ever before are not forced to pilot new processes or to use a previous design for a new application. They have become more efficient, hopefully more productive, and have been relieved of some of the routine business to allow more time for the creative aspect of engineering. The advent of the electronic office is a step in this direction, replacing time consuming manual transactions with more efficient automated ones.

The chemical engineering educational system has been profoundly affected by computers, however. Consider the impact on the curriculum, for example. Numerical analysis, computer programming, and other computer related courses have become standard offerings. Process control and process design courses are taught much differently today in many universities than twenty-five years ago. Computers have allowed us to expand our theoretical understanding of chemical engineering. They have allowed us to conduct research studies that could not have even been done before.

As a profession where would we be without the influence of computers? Because we are more productive, our salaries are probably higher. Without computers engineers would probably be performing more mundane or routine tasks. This could mean that chemical engineering employment might be higher today than it is - but would the work be as interesting, or challenging?

What about the chemical industry? Computers have enabled us to design more reliably, more economically. Without them plants would be more expensive to build and maintain, more energy inefficient, less pollution-free. Process startups are more efficient, and process development time and costs are significantly less. New technology is more easily evaluated and introduced. Simulation technology has both reduced the required effort and increased the efficiency of process research and development.

The impact on society is similar. Products that we have would probably be more expensive. Some products might not even exist because the plants to manufacture them would not be economical. Our environment would likely be more polluted. As mentioned before, chemical plants would be less energy efficient, but technological advances would have been constrained so that overall energy use per capita may have been less - the net impact on energy is not clear.

In summary, computers have had and will continue to have a significant impact on chemical engineering. The advances have been far reaching and have affected not only the profession of chemical engineering, but the chemical industry and society as well.

ACKNOWLEDGEMENTS

The author wishes to acknowledge the assistance and counsel of a number of people whose inputs and advice were critical to the development of this paper. These are: J. R. Deam, H. H. Chien, G. W. Hopwood, R. C. Morris, J. D. Nutter, R. E. Otto, and E. M. Rosen at Monsanto; J. V. Heck, IBM, R. S. Mah, North-

western; R. L. Motard, Washington University;
J. L. Robertson, Exxon; W. D. Seider, University of Pennsylvania; A. W. Westerberg,
Carnegie-Mellon; and T. J. Williams, Purdue.
Much of what is in this paper has come from
private communications with these people. A
considerable amount of information on the developments in computer hardware and software
has come from the IBM Corporation through the
generous assistance of J. V. Heck.

HISTORY OF DISTILLATION CONTROL

P.S. Buckley ■ Engineering Department, E.I. du Pont de Nemours, Inc., Wilmington, DE 19898

The history of distillation control is discussed in terms of developments in column design methods, column tray and auxiliary design, instrumentation, control system design methods, and control techniques.

INTRODUCTION

History is normally approached in a chronological fashion, starting at an earlier date and proceeding step wise and logically to a later date. To discuss distillation control in this manner is difficult. Most important developments have been evolutionary, occurring gradually over a period of time. Consequently, we will be discussing this subject mostly in terms of eras, rather than specific dates. Further, the time period of most interest for distillaton control history is very short - about thirty-five years. As every historian knows, the closest one comes to the present, the more difficult it is to achieve a valid perspective. The bias becomes worse when the historian himself has been a participant in the events to be discussed.

WHAT IS CONTROL?

A chemical plant or refinery must produce a product or products which meet certain quality specifications. Preferably, there should be enough inventory to ship upon receipt of orders, although this is not always feasible, and some orders call for shipping a certain amount continuously or intermittently over a period of time. Plant operation must meet production requirements while observing certain constraints of safety, environmental protection, and limits of efficient operation. We do not sell flow rates, liquid levels, temperatures, pressures, etc. Contrary to popular belief, these variables need not be held constant, but should mostly be manipulated to achieve the following operational objectives:

1. Material balance control - an overall plant material balance must be maintained. Production rate must on the average equal rate of sales. Flow rate changes

2. Product quality control - final product or products must meet sales specifications. But product quality needs to be constant at only one point in a process - final inventory.

3. Constraints - if there is a serious malfunction of equipment, interlocks may shut the plant down. Otherwise, override controls may nudge a process away from excessive pressure or temperature, excessively high or

low liquid levels, excessive pollutants in waste streams, etc.

An important implication of the preceeding is that for optimum operation, one must let all variables vary somewhat except final product quality. And sometimes, even that may vary within prescribed limits.

Classical control theory emphasizes (a) rapid response to setpoint changes and (b) rapid return to setpoint in the face of disturbances. These performance requirements really apply only to product quality controls and to some constraint controls. Some of the successes of computer control have resulted from restoring control flexibility which was originally lost by excessive reliance on fixed setpoint control of individual variables.

COLUMN DESIGN METHODS

Other speakers are going to cover this in detail, so I will limit myself to a few simple points.

In the 1930's and 1940's, what chemical engineers call "unit operations theory" made tremendous strides. The objectives were twofold: (1) to ensure that design of piping, heat exchangers, distillation columns, etc, would have at least flow sheet capacity, and (2) that this equipment would be as little overdesigned as possible to minimize capital investment. For distillation columns, the McCabe-Thiele procedure for estimating the required number of theoretical trays was published in 1925. It did not however, (at least in my experience) achieve a large usage until the early 1940's. During World War II a group of us trying to increase production rates in a heavy chemicals plant found that existing columns designed before 1930 occasionally had tremendous overdesign factors - they were ten to fifteen times larger than necessary.

As well as I can recall, by 1948 we were using a safety factor of two for designing distillation columns and

heat exchangers, and 25-50% for piping.

Today overdesign safety factors are very small. Columns are designed to run much closer to flooding, and thermosyphon reboilers often operate close to choked-flow instability. Older columns with more safety factor in design and particularly with bubble cap trays, were easier to control.

COLUMN TRAY AND AUXILIARY DESIGN

Tray Design

About 1950 we began to design columns with sieve trays instead of bubble-cap trays. For a given load, a sieve tray column is smaller and cheaper. But sieve tray columns have more limited turndown, about 2 to 1 instead of about 8 to 1. In addition they are subject to weeping and dumping. These characteristics, together with tighter design practices, increase control problems.

More recently, valve trays have become popular. They have more turndown capabilities than sieve trays, say 4:1, but have the same potential for weeping and dumping. In addition they commonly have another problem: "inverse response".

As noted by Rijnsdorp [1], with an increase in boilup, some columns show a momentary increase in internal reflux, followed eventually by a decrease in internal reflux. This is termed "inverse response" since it causes a temporary increase in low boiler composition at the bottom of the column, followed by an eventual decrease. The mechanism was elucidated by Buckley, Cox and Rollins [2] who found that most sieve tray columns demonstrated this effect at low boilup rates, while most valve tray columns demonstrated it over the entire turndown. Thistlethwaite [3] also studied inverse response and worked out more details. Composition control via boilup at the base of a column is vastly more difficult when a column is afflicted with inverse response. Momentarily, the controller seems to be hooked up backwards. For the control engineer this can be even worse than dead time.

Reboiler Design

Thermosyphon reboilers, particularly of the vertical type, are popular because of low cost. Several kinds of dynamic problems, however, can result from their use (4). One is choked flow instability (5). This occurs when the reboiler is too snugly sized with excessive heat flux.

Another problem is "swell". At low heat loads there is little vapor volume in the tubes; at high heat loads there is much more vapor in the tubes. Hence, an increase in steam flow will cause liquid to be displaced into the column base, causing a temporary increase in base level. If base level is controlled by steam flow, the controller momentarily acts like it is hooked up backwards. When "swell" is accompanied by inverse response of reflux, controlling base level via heating medium may be impossible.

Recent trends in thermosyphon design seem to have had at least one beneficial effect - critical T is rarely a problem. Prior to 1950, it could be observed fairly often. This improvement apparently is due to much higher circulating rates.

Flooded Reboilers

Particularly with larger columns it is common to use steam condensate pots with level controllers instead of traps. In recent years some reboilers have been designed to run partially flooded on the steam side.

The large steam control valve may be replaced by a much smaller valve for steam condensate. Also, it is usually easier to measure condensate flow rather than steam flow. Flooded reboilers are, however, more sluggish than the nonflooded ones, typically by a factor of 10 or more.

Column Base Designs

The size of the column base, the design of the internals (if any), and the topological relationship to the reboiler are all important factors in column control. Column base holdups may range from a few seconds to many hours (in terms of bottom product flow rate). If there are no intermediate holdups between columns or process steps, the column base must serve as surge capacity for the next step.

Originally, column bases were very simple in design but in recent years elaborate internal baffle schemes have often been used. The objective is to achieve one more stage of separation. Many of these schemes make column operation more difficult, and limit the use of the hold-up volume as surge capacity. Overall, their economics is questionable; increasing the number of trays in a column by one is relatively cheap.

Overhead Design

In the chemical industry, particularly for smaller columns, it is common to use gravity flow reflux and vertical, coolant-in-shell condensers. Instrumentation and controls can be much simpler than with pumped-back reflux. Regardless, however, of condenser design, gravity flow reflux systems have a dynamics problem if distillate is the controlled flow and reflux is the difference flow. This is the well-known problem of "reflux cycle". Its mechanism and several corrective measures are discussed in a paper by Buckley (6).

When horizontal, coolant-in-tube condensers are used, control is sometimes simplified by running the condensers partially flooded. This technique is especially useful when top product is a vapor. Flooded condensers seem to be becoming more popular, but like flooded reboilers, are more sluggish. For both, however, we have worked out mathematical models which enable us to calculate the necessary derivative compensation.

INSTRUMENTATION

Before automatic controls and automatic valves become prevalent, control was achieved by having operators observe local gages and adjust hand valves to control process variables to desired values. According to stories told me by old timers, some applications required such careful attention and frequent adjustment that an operator would be kept at a single gage and hand valve. To help keep him alert, he was provided with a one-legged stool.

Gradually, this practice diminished as manufacturers developed automatic controls in which measurement, controller, and valve were combined all in one package. Sometimes the valve itself, usually powered by air and a spring and a diaphragm actuator, was a separate entity. Slowly, the practice of separating measurement, controller, and valve evolved.

The development of measurement devices which could transmit pneumatic or electric signals to a remote location was well underway by 1940. This permitted the use of central control rooms and together with more automatic controls, a reduction in the number of operators.

By 1945 the chief instruments available for distillation control were

o Flow - orifice with mercury manometer plus pneumatic transission, or rotameters

o Temperature - thermocouple or mercury-in-bulb thermometers

o Pressure - bourdon tube plus pneumatic transmission

o Control valve - air-operated with spring - and - diaphragm operator, or electric motor operator. Most valves, except for small ones, were of the double-seat variety. Small ones were single-seat.

Valve positioners occasionally were used.

By 1950 the picture had begun to change dramatically. The so-called force balance principle began to be used by instrument designers for pneumatic process variable transmitters and controllers. This feature together with higher capacity pneumatic pilots, provides more sensitivity, freedom from drift, and speed of response. Mercury-type flow transmitters gave way to "dry" transmitters and mercury-in-bulb temperature transmitters were replaced by gas-filled bulb transmitters.

Piston-operated valves with integral valve positioners appeared on the market. By making single-seated valves in larger sizes practical, they permitted less expensive valves for a given flow requirement. They also provided increased sensitivity and speed of response.

New pneumatic controllers appeared with external reset feedback, permitting the use of antireset wind-up schemes. One vendor put on the market about 1952 an extensive line of pneumatic computing relays, which permitted addition, subtraction, multiplication by a constant, high- and low-signal selection, and so-called impulse feedforward. Really decent pneumatic devices for multiplying or dividing two signals did not appear until the mid 1960's. Taken together, these various devices permitted on-line calculation of heat and material balances. Calculation of internal reflux from external reflux and two temperatures also became practical. And simple pressure-compensated temperature schemes became feasible.

The improved Δp transmitters, originally developed for orifice flow measurements, permitted sensitive specific gravity and density measurements. They could also be used to measure column Δp, increasingly important for columns designed to run closer to flooding.

Parallel to developments in conventional instruments there was a much slower development in on-line analyzers. In the late 1940's, Monsanto developed an automatic refractometer which was used in a styrene monomer distillation train. In the early 1950's, I can remember seeing several prototypes of on-line infrared and optical analyzers. Problems with reliability and with sample system design plus high cost severely limited their applications. pH measurements, on the other hand, expanded more rapidly; this came about from improved cell design, better cable insulation, and high input impedence electronic circuitry.

The first analog electronic controls appeared in the mid 1950's. They had good sensitivity and speed of response but due to the use of vacuum tubes, had somewhat high maintenance. In addition, they were far less flexible than pneumatics. Switching to solid state circuitry at a later date improved reliability but flexibility still suffered. It was not until the early 1970's that one manufacturer produced an electronic line with all of the computation and logic function available in existing pneumatics. In the meantime, the cost competitiveness of pneumatics was improved by a switch from metallic transmission tubing to black polyethylene tubing outdoors and soft vinyl tubing in the control room.

In the late 1970's a potentially momentous change got underway - the introduction of so called "distributed" digital control systems. By using a number of independent microprocessors, manufactures were able to "distribute" the control room hardware and achieve better reliability than would be obtained with a single, large time-shared computer (avoid "all eggs in one basket"). Early versions were quite inflexible compared to the best pneumatic and electronic hardware but this situation has now changed. Some of the newest systems have more flexiblity than any analog system and lend themselves well to the design of advanced control systems. Transmitters and control valves in the field may usually be either pneumatic or electronic.

No discussion of instrumentation is complete without some mention of computers. The first process control computer was put on the market in the early 1950's. A well-publicized application was a Texaco refinery, where some optimization was performed. Another optimization application of that era was a Monsanto ammonia plant at Luling, LA. Other machines came on the market in the next few years. But all of the early machines suffered from lack of reliability, slow operation, and insufficient memory. In the 1960's and 1970's applications were chiefly as data loggers and supervisory controls. Production personnel found that CRT displays, computer memory, and computer printouts were a big improvement over conventional gages and recorders. As memory and speed of computers have increased and as the expertise of users has improved, we have seen a resurgence of interest in optimization.

The latest development in instrumentation will begin in mid - 1983 when one vendor will put on the market a line of "smart" measurements. These will feature built-in microprocessors to permit remote automatic calibration, zeroing, and interrogation. These features, together with reduced drift and sensitivity to ambient conditions, and with more accuracy and linearity, will improve plant operation. They will also facilitate and reduce maintenance. Eventual extension to instruments with multiple functions such as temperature and pressure is foreseen. Present indications are that within the next several years a number of manufactures will put "smart" instruments on the market.

CONTROL SYSTEM DESIGN METHODS

Prior to 1948, distillation control systems as well as other process control systems were designed by what might be called the "instrumentation" method. This is a qualitative approach based principally on past practice and intuition. Many "how to do it" papers have appeared in the literature. A good summary of the state-of-the-art in 1948 is given by Boyd (7).

In the period 1948-1952 there appeared a large number of books based on control technology developed for military purposes during World War II. They commonly had something in the title about "servomechanisms theory". Two of the best known are by Brown and Campbell (8) and Chestnut and Mayer (9). A war time text by Oldenbourg and Sartorius was translated from German into English (10) and became well known. Although the theory required significant modifications to adapt it to process control, for the first time it permitted us to analyze many process control situations quantitatively. The technique involves converting process equations, including differential equations, into linear perturbation equations. By perturbations we mean small changes around an average value. Linear differential equations with constant coefficients are easily solved by use of the Laplace transformation. It is usually easy to combine a number of system equations into one. This equation, which relates output changes to input changes, is called the "transfer function". Time domain solutions are readily calculated, and if the uncontrolled system is stable, the transfer function may be expressed in terms of frequency response.

The transfer function approach permits convenient study of system stability, permits calculation of controller gain and reset time, and permits calculation of system speed of response to various forcing functions. In the precomputer era, these calculations were made with a slide rule and several graphical aids. Today the transfer function techniques are still useful but the calculations may usually be made more rapidly with a programmable calculator or a computer. Early

applications were in flow, pressure, and liquid level systems.

Throughout the 1950's and 1960's the transfer function approach was used extensively for piping, mixing vessels, heat exchangers, extruders, weigh belts, and occasionally distillation columns. Two books which appeared in the 1950's illustrated the application of transfer function techniques to process control (11) (12).

The relatively simple modeling techniques developed earlier were of limited use for systems describable by either partial differential equations or by a large number of ordinary differential equations. Gould (13) in 1967 explored some more advanced modeling techniques for such systems.

The limitations of the simple transfer function approach for distillation columns led some workers to try another approach - computer simulation. The work of T. J. Williams and associates (14) (15) in applying this approach is well known. Basically simulation consists of writing all of the differential equations for the uncontrolled system and putting them on the computer. Various control loops can then be added, and after empirically tuning the controllers, the entire system response may be obtained. In some cases real controllers, not simulated ones, have been connected to the computer.

In 1960, Rippin and Lamb (16) presented a combined transfer function - simulation approach to distillation control. The computer simulation was used to develop open loop transfer functions. Then standard frequency response methods were used to calculate controller gain and reset as well as desired feedforward compensation. In our own work we have found this combination approach to be quite fruitful. .

As the speed and memory of technical computers increased, distillation models have become more detailed and elaborate. Tolliver and Waggoner

(17) have presented an excellent discussion of this subject.

Since about 1970 these has been increasing interest in newer approaches to control. The incentives have been twofold:

1. Since flow sheet data is often incomplete or not exact, it is desirable to measure process characteristics of the real plant after start-up. Furthermore, process dynamics change as operating conditions change. The increasing availability of on-line computers permits the use of "identification" techniques. Information thus obtained may be used to return controller either manually or automatically.

2. Many processes are plagued by interactions. There are a variety of techniques for dealing with this which may be lumped under the heading "Modern Control Theory". That aspect relating to time-optional control will likely be helpful in future design of batch distillation columns. Implementation of this technology usually requires an on-line computer.

These newer techniques require a lot of training beyond that required for transfer function techniques, and industrial applications have so far been limited. There is no question, however, that there exist ample incentives and opportunities for this newer control technology.

An interesting trend in recent years has been toward the development of standardized programs for control system design. This is commonly call CAD - Computer Aided Design. Although originally aimed at larger machines, many useful programs exist for small computers and even programmable calculators. The portability and permanent memory features of many of the latter render them very useful for taking into a plant or maintenance shop.

PROCESS CONTROL TECHNIQUES

In the 1945 there existed no overall process control strategy. As one process design manager put it, "instruments are sprinkled on the flow sheet like ornaments on a Christmas tree". Not until 1964 was there published an overall strategy for laying out operational controls from one end of a process to the other (18). This strategy, as mentioned earlier, has three main facts: Material balance control, product quality control, and constraints. What kind of controls do we use to implement these?

Material balance controls are traditionally averaging level and averaging pressure controls. Although the basic concept dates back to 1937, averaging controls are still widely misunderstood. The original theory was expanded in 1964 (18) and again quite recently (19). To minimize required inventory and achieve maximum flow smoothing we use quasi-linear or nonlinear PI controllers cascaded to flow controllers. If we can use level control of a column base to maniuplate bottom product flow and level control of the overhead receiver to adjust top product flow, then we can usually avoid the use of surge tanks between columns or other process steps. This saves both fixed investment and working capitol.

Product quality controllers are usually also of the PI type. Overhead composition is usually controlled by manipulating reflux and base composition by manipulating heat flow to the reboiler. It is also increasingly common to provide feedforward compensation for feed rate changes. This minimizes swings in boilup and reflux, and in top and bottom compositions.

One of the most important developments in control techniques is that of variable configuration and variable structure controls. While it is much more common to use fixed configuration and fixed structures, in reality either or both should change as process conditions change. For example, the steam valve for a distillation column reboiler may, depending on circumstances, respond to controllers for

    Steam flow rate
    Column Δp
    Column pressure
    Base temperature
    Column feed rate
    Column base level
    Column bottom product rate

The seven variables listed may also exert control on five or six other valves. We call this "variable configuration".

For level control, the quasi-linear structure referred to earlier consists of PI control normally, but proportional - only control when level is too high or too low. We call this "variable structure".

To accomplish this automatically, it was proposed in 1965 (20) to use a type of control called "overrides". In ensuing years a number of other publications (21) (22) (23) described the applications to distillation. The math and theory are presented in a two-part article (24).

Fairly recently these has been growing an interest in "self-tuning regulators". One of the most prolific contributors has been Åström (25) in Sweden. So far, applications to industrial distillation columns are not known, but thay have been tried on pilot columns.

INFLUENCE ON DISTILLATION CONTROL

Some years ago I was present when one of our most experienced distillaton experts was asked, "What is the best way to control a column?" His answer: "gently". When approaching column control one should keep in mind that conditions in a column cannot be changed rapidly. Rapid changes in boilup or rate of vapor removal, for example, may cause momentary flooding or dumping. The developments since 1945 have therefore not been aimed at more rapid control but rather at smoother operation with better separation at lower cost.

If we now look back at 1945 and move forward we can see how the various developments previously discussed have interacted to influence distillation control.

1. Switch to sieve and valve trays and tighter column design

    These have encouraged the user of minimum and maximum boilup overrides. A high Δp override on steam is now universally employed by some engineering organizations. We also frequently provide maximum and minimum overrides for reflux.

2. Tighter design of heat exchangers, especially reboilers

    To prevent choked flow instability we sometimes provide maximum steam flow limiters.

3. Increased use of side-draw columns

    One side-draw column is cheaper than two conventional columns with two takeoffs. Turndown is less, however, and the column is harder to control (26) (27) (28). Such columns usually require more overrides, particularly to guarantee adequate reflux below the side draw or draws.

4. Energy Crunch of 1973

    This has had two major effects:

    4.1 Increased energy conservation of conventional columns. After improved insulation, one of the big items has become double-ended composition control. A column runs more efficiently and has more capacity if composition is controlled at both ends.

Another consequence has been more interest in composition measurements instead of just temperature. Temperature is a nonspecific composition measurement except for a binary system at constant pressure. This, together with the common practice of having temperature control of only one end of a column, usually leads production personnel to run a column with excess boilup and reflux to ensure meeting or exceeding product purity specifications. This wastes steam and limits column capacity. Experience indicates that 10-30% savings are often archievable. To get an idea of the numbers involved, consider a column which uses 1,000 pph. of steam. If we can save 100 pph, if average operating time is 8,000 hours/year, and if steam costs $5/1,000 lbs, then annual savings are $4,000. This means we can invest $8,000-$12,000 in improved controls to achieve the desired steam saving.

4.2   Heat Recovery Schemes

Much larger energy savings are often possible with energy recovery schemes, usually involving the condensation of vapor from one column in the reboiler of another column (29). Our experience has been that designing controls for energy recovery schemes is much more difficult and time consuming than for conventional columns. This extra cost, however, is usually a small fraction of savings.

An interesting feature of some heat recovery schemes

is that the pressure in the column supplying heat is allowed to float. Without going into details I will simply say that we have found that this simplifies process design (usually no auxiliary condensers are needed) and simplifies instrumentation and control (no pressure controller is needed). Since maximum operating pressure occurs when boilup is a maximum there is no problem with flooding.

There are other problems, however. For example, the feed valve to a column with floating pressure has a much larger Δp variation with feed rate than is commonly encountered. And the column bottom product valve has a low Δp at low rates and a high Δp at high rates. To obtain reasonably constant stability, one may need to provide "flow characteristic compensation" (30). This need is increased by a trend toward lower Δp for valves in pumped systems.

There are some differences in control requirements between the chemical and petroleum industries. Tolliver and Waggoner (17) have presented an excellent discussion of this subject.

Literature

There are now several texts on distillation control. The one with the most complete technical treatment is that by Rademaker, Rijnsdorp, and Maarleveld (31). Another, by Shinskey (32) is notable for its treatment of energy conservation. A third, by Nisenfeld and Seemann (33), has a well-organized treatment of distillation fundamentals and types of columns. Neither of these last two books deals with control in a technical sense - there is no stability theory, Laplace transforms, or frequency response. But each contains many gems of practical details about column control.

CONCLUDING REMARKS

From the preceeding we can see that since 1945 we have moved from generously designed columns with bubble cap trays and simple controls to tightly designed columns with sieve or valve trays and with more sophisticated controls. For new design projects with difficult applications it may be appropriate in some cases to consider going back to bubble cap trays and larger design safety factors to avoid the necessity of very complex and perhaps touchy controls.

The literature on distillation control, particularly that from academic contributors, emphasizes product composition control. Our experience has been that many operating difficulties, in addition to those due to tight design, are due to improperly designed column bases and auxiliaries, including piping. In my own experience, the single factor which contributes most to low cost operation and good composition control has been properly designed averaging liquid level controls.

As of mid - 1983, lack of adequate product quality measurements is probably the Achilles heel of distillation control. With adequate quality measurements and composition control of each product stream, the next problem area will be that of interactions.

Although a great deal of progress has been made in quantitative design of control systems, I would say that today we are about where distillation design was in 1948.

REFERENCES

1. Rijnsdorp, J. E., Birmingham University Chemical Engineer, vol 12, 1961, pp 5-14.

2. Buckley, P. S., R. K. Cox and D. L. Rollins "Inverse Response in a Distillation Column", CEP vol 71, No. 6 (83-84), July 1975.

3. Thistlethwaite, E. A., "Analysis of Inverse Response Behavior in Distillation Columns", M. S. Thesis, Department of Chemical Engineering, Louisiana State University, 1980.

4. Buckley, P. S., "Material Balance Control in Distillation Columns", paper presented at AIChE Workshop, Tampa, FL, November 1974.

5. Shellene, K. R., C. V. Stempling, N. H. Snyder and D. M. Church, "Experimental Study of a Vertical Thermosyphon Reboiler", presented at Ninth National Heat Transfer Conference AIChE-ASME, Seattle, WA, August 1967.

6. Buckley, P. S., "Reflux Cycle in Distillation Columns", presented at IFAC Conference, London 1966.

7. Boyd, D. M., "Fractionation Instrumentation and Control", Petr Ref, October and November 1948.

8. Brown, G. S. and D. P. Campbell, "Principles of Servomechanisms", John Wiley, NYC 1948.

9. Chestnut, H. and R. W. Mayer, "Servomechanisms and Regulating System Design", vols I and II, John Wiley, NYC, 1951.

10. Oldenbourg, R. C. and H. Sartorius, "The Dynamics of Automatic Controls", ASME, NYC, 1948.

11. Campbell, D. P., "Process Dynamics", John Wiley, NYC, 1958.

12. Young, A. J. (ed), "Plant and Process Dynamic Characteristics", Academic Press, NYC, 1957.

13. Gould, L. A., "Chemical Process Control", Addison-Wesley, Reading, MA, 1969.

14. Williams, T. J., B. T. Harnett, and A. Rose, Ind & Eng Chem, 48, No. 6, 1008-1019 (1956).

15. Williams, T. J., "Distillation Column Control Systems", Proceedings of 12th Annual Chem & Petr Ind Symposium, Houston, April 1971.

16. Rippin, D. W. T. and D. E. Lamb, "A Theoretical Study of the Dynamics and Control of Binary Distillation", Bulletin from University of Delaware, 1960.

17. Tolliver, T. L. and R. C. Waggoner, "Distillation Column Control", ISA, 1980, paper No. C. I. 80-508.

18. Buckley, P. S., "Techniques of Process Control", John Wiley, NYC, 1964.

19. Buckley, P. S., "Recent Advances in Averaging Level Control", presented at ISA meeting, Houston, April 1983.

20. Buckley, P. S., "Override Controls for Distillation Columns", INTECH, August 1968, pp 51-58.

21. Buckley, P. S. and R. K. Cox, "New Developments in Overrides for Distillation Columns", ISA TRANS, vol 10, No. 4, 1971, pp 386-394.

22. Buckley, P. S., "Protective Controls for Sidestream Drawoff Columns", presented at AIChE Meeting, New Orleans, March 1969.

23. Cox, R. K., "Some Practical Considerations in the Application of Overrides", presented at ISA Symposium, St. Louis, MO, April 1973.

24. Buckley, P. S., "Designing Override and Feedforward Controls", Control Engineering, Part 1, August 1971, pp 48-51, Part 2, October 1971, pp 82-85.

25. Åstrom, K. J., "Self-Tuning Regulators", Report LUTFD2/ (TRFT-7177)/1-068 (1979), Lund Institute of Technology, Sweden.

26. Buckley, P. S., "Control for Sidestream Drawoff Columns", CEP, vol 65, No. 5, pp 45-51, May 1969.

27. Doukas, N. and W. L. Luyben, "Control of Sidestream Columns Separating Termary Mixtures", INTECH pp 43-48, June 1973.

28. Luyben, W. L., "10 Schemes to Control Distillation Columns with Sidestream Drawoff", ISA Journal, pp 37-42, July 1966.

29. Buckley, P. S., "Control of Heat - Integrated Distillation Columns" presented at Engineering Foundation Conference, Sea Island, GA, January 1981.

30. Buckley, P. S., "Optimum Control Valves for Pumped Systems", presented at Texas A&M Symposium, January 1982.

31. Rademaker, O., J. E. Rijnsdorp, and A. Maarleveld, "Dynamics and Control of Continuous Distillation Columns", Elsevier, NYC 1975.

32. Shinskey, F. G., "Distillation Control", McGraw-Hill, NYC 1977.

33. Nisenfeld, A. E. and R. C. Seemann, "Distillation Columns", ISA Research, Triangle Park, NC, 1981.

COMPUTERS IN CHEMICAL ENGINEERING— CHALLENGES AND CONSTRAINTS

R.W.H. Sargent ■ Imperial College, London, England

Computer science and technology continue to develop at an ever-increasing pace. As chemical engineers in industry and in the universities come to terms with these developments, new methods of working and new capabilities emerge. An attempt is made to pick out the significant trends, in the light of both current process systems research and current developments in computer science, and to discuss the implications for the future everyday work of the chemical engineer and the consequential changes in his education.

Computers and computing are less than half as old as the A.I.Ch.E., but their influence has become all-pervasive. Space-invaders have now invaded the home, and schoolchildren clamour for home computers whose facilities would have been beyond the dreams of most university departments only a few years ago. Word-processing systems have invaded the office, revolutionizing procedures, automating filing systems, and providing instant communication. Computer suppliers, computer bureaux and software houses proliferate, offering packages for every kind of calculation or management task.

Chemical engineers have certainly not been behind in exploiting the benefits of computing, and our ways of working have indeed changed very significantly over the past thirty years. But the pace is quickening, and the last few years have seen a veritable explosion of new ideas, new hardware and new techniques. It is easy to allow oneself to be carried along on the tide of enthusiasm and euphoria, so it is perhaps timely to stand back a little, and re-examine our own problems and aspirations in the light of these new possibilities.

Imperial College, London.
October 1983

Computers and Design

The influence of computer developments on the various stages of the design process was the theme of the recent FOCAPD-83 Conference, and the major trends are discussed in the present author's introductory paper to that conference (Sargent 1983a).

The development of a process from the initial idea to the commissioning of the plant is a large collaborative exercise, involving many different groups of people from different disciplines, coming from different departments within the organization, or even from different companies. Thus a major contribution to the efficiency of the enterprise can be made simply by the computerization of office filing systems and office procedures, providing easier accessibility to relevant information, easier updating of this information, and easier communication.

Many of the individual steps in the design process are of course already computerized, and an impressive array of packages is available to assist the designer. With recent developments of specialized hardware and software, we now have cost-effective means of entering, storing, manipulating and displaying, both on a screen and in hard-copy, all kinds of alpha-numeric and graphical data. It is thus possible to think in terms of a work-station for the design engineer which enables him to work with the computer, directly and interactively, on every aspect of the design problem.

Unfortunately, most of the computer programs have been developed independently as stand-alone packages, and the engineer is faced with the time-consuming and error-prone process of transcribing the output data from one package into the required form for input to the next - and there are at present some important linking steps which are not yet computerized. Attempting to prescribe stand-ard interfaces between the various design operations is a hopeless task, and a change of view point is necessary. Instead of concentrating on programs which execute given tasks, and thinking about the resulting input and output requirements, we need to focus attention on the data, which represents the current state of the design - programs and input from the designer are just means of generating and modifying this data. If all the project data is stored systematically in a properly structured data-base, then it is easy to provide an interface for any package operating on any portion of the data. It also simplifies the task of selecting relevant data for different aspects of the design, and indeed for ensuring that each piece of data is available only to those who have the appro-priate authorization.

With today's data-base technology it would in principle be possible to construct an integrated data-base covering every phase of the whole development project, each group creating, using or modifying the data as it makes its contribution. However the amount of data involved would be quite stupendous, and to make the management of it a practical proposition, the data-base would have to be highly structured. The structure would have to reflect the structure of the project management organization itself, and be capable of evolving with the project. To obtain efficient access, and to maintain consistency as changes are made, it would also have to reflect the structure of the process with its various functional relationships. We know all too little about the construction and manage-ment of data-bases with such multi-layered structures, and there is a real challenge here for both engineers and computer scientists.

In the meantime the problem is being tackled on a less comprehensive scale, with the development of data-base systems and associated design packages for various individual phases of design and construction. Several of these systems were described at the recent Eurochem 83 Conference, and it is clear that we shall soon be faced with the problem of interfacing these systems to each other. However the cross-linking of data-bases should be a less daunting task than the interfacing of individual programs, and will eventually be a natural step in the creation of more comprehensive structured data-bases.

In recent years the process industries have become more competitive. Profit margins are smaller, and greater social consciousness forces closer attention to safety and environmental impact. These trends demand better, and more predictable, process performance, as well as greater efficiency in the design process itself, and this in turn implies the use of more detailed and more realistic predictive models, and a great increase in computing requirements. Thus, in spite of the impressive increases in computing power over the years, there continue to be areas where the eningeer's scope is limited by computing capability.

In fact it is not only a matter of computational speed. The large complex models we wish to handle today pose serious challen-ges to existing numerical techniques, and it is even more important to develop robust methods which guarantee a solution under a useful range of conditions. The robustness of existing methods can often be greatly improved by careful implementation and attention to proper scaling of the variables and functions involved. Numerical analysts are also constantly improving the underlying methods and producing new algorithms. The engineer can also help in the way he form-ulates his models, for there are usually many alternative ways of writing the describing equations for a particular model, and some are much better for numerical solution than others, Of course, in principle all the alternatives can be generated from any given formulation by appropriate manipulations, and we might hope to hand over the task of choosing the best form to the computer. Unfortunately, present automatic algebraic manipulation packages are hopelessly inefficient, both in storage and processing time, though there are signs that things can be dramatically improved. In any case however, the number of alternatives is usually vast, and it is not clear how to formulate appropriate criteria for selecting the best among them. So until the computer scientists and mathematicians have made more progress on these problems, we must rely on the engineer developing a good numerical sense to guide his formulation of the model.

The real hope for improving robustness is in fact to exploit the engineer's general knowledge of the physical behaviour of his systems. In the early days of chemical engineering, this did indeed play an important role in developing effective numerical and graphical solution methods, but for dealing with the infinite variety of present-day models, we need a way of using the engineer's physical intuition without requiring him to devize a specific algorithm for each problem.

Of course many of the physical characteristics are embedded in the mechanistic models of the various basic physical phenomena we use in writing down our describing equations. However, in the complete system these elementary processes often interact in complicated ways, and the engineer's intuition consists of recognizing which are the dominant processes, or of a broad knowledge of the general form of the resultant behaviour of particular structures. In either case, this knowledge can usually be embodied in a simplified or "short-cut" model. There is therefore the new and challenging problem of devizing general algorithms which exploit both simple and complete models to find solutions to a specified accuracy.

The same pressures for improved plant performance have led to the recognition that a design based on steady-state operation to meet a single set of closely defined specifications simply will not do, and the emphasis has moved to the design of plants which will operate satisfactorily over a range of conditions. This includes the problem of rapid change-over from one set of conditions to another, and there is a natural extension to operation of mixed batch and continuous plant, and other multi-purpose plant, with their associated scheduling problems. There is also a growing recognition that, as plants become more integrated, with less intermediate storage, the designer must also give closer consideration to the question of operability and control.

The transition from feasible to optimal design is easy, for we now have powerful nonlinear programming algorithms, even for large-scale problems. However ensuring feasible operation over a range of conditions

makes the problem infinite-dimensional, and dealing rigorously with uncertainty again introduces infinite-dimensional distribution functions. Taking account of the dynamics creates an optimal control problem, while changing the configuration of the plant to meet changing conditions makes it a combinatorial problem. In all these extensions the numerical problems are formidable, and the costs of computing are a real constraint. However progress is being made, and Grossman and Morari (1983) gave an excellent exposition at FOCAPD 83 of the problems, and of the techniques being developed to handle them.

Design is not just a matter of data-handling and calculations. Some of the most vital decisions on the design concern qualitative choices, such as the choice of the processes and equipment to be used, and many more decisions involve weighing quantitative measures of performance, like costs and yields, against more qualitative considerations, like operability and safety.

For some years now chemical engineers have been grappling with the problem of dealing with these qualitative issues within computer-based design packages, though most of this work has been in the area of "process synthesis", the early phase of design concerned with choice of the chemical route from raw materials to products, choice of separation processes, and development of the flowsheet, involving sequencing of unit operations, materials recycling, and energy integration.

The first problem to be faced is a representation of the qualitative structure so that it can be stored in the computer - in effect a definition of the basic building blocks and the rules for their association, which together provide a systematic means of generating all possible structures. For example, in choosing the chemical route, the building blocks are chemical compounds, and the association rules are the laws governing chemical reactions. In the case of hazard analysis, the blocks are "events" or "situations", linked by cause and effect relationships.

The second problem is a means of evaluating the structures generated. For physical structures, the technological features may be sufficiently well defined to make it possible to assign a cost, or at least a good estimate of the cost. For

other features it may also be possible to
formulate a quantitative performance criterion;
for example in energy considerations
thermodynamic criteria can be used, while
control theory can provide an appropriate
measure for operability. In other cases, the
most we can hope for is some kind of ranking
of the different solutions. Of course in most
situations the design must be considered in
relation to a number of different criteria,
and the question then arises of whether they
can be simultaneously satisfied, and if not
what relative weighting or priority order
should be used. In the last resort, the
system must present the feasible choices to
the designer to allow him to choose, but
this is only a practical solution if the
choices presented can be narrowed to a
reasonable number.

With the representation and evaluation
problems resolved, mathematical techniques
can be brought to bear to direct the search
for the optimal or preferred solutions.
However, even with good selection and rejec-
tion rules, the number of cases to be examined
is usually huge, and we again find our scope
limited by computational requirements. In
the later stages, much of the evaluation will
involve detailed design calculations, or
perhaps detailed dynamic simulations to
resolve issues of safety or operability.
However the number of cases to be examined in
such detail must be severely limited, and
again we must appeal to the intuition of the
designer to provide much simpler models for
evaluation in the earlier stages, or even
heuristic rules for making choices without the
need for computations.

A computer-based system which relies
completely on a human expert to provide
knowledge in the form of facts and their
logical connections, and which uses this
"knowledge-base" to answer queries on the
area covered, is called an "expert system".
Systems like this have been produced in such
diverse fields as medical diagnosis, chemical
synthesis, hazard analysis, aid in prospecting
for minerals, and advice on the appropriate
numerical method to solve a given problem.

The construction of such systems has been
made much easier by the development of
special logic-based programming languages,
which represent an entirely new approach to
computer programming. These languages also
provide a more natural tool for dealing with
structured data-bases, and have the great

advantage that it is possible to describe a
problem without at the same time having to
prescribe a solution method. This latter
property is clearly important for situations
such as occur in design, where the problem
description gradually evolves, and is then
used for a variety of purposes.

Clearly these languages are an ideal tool
for building comprehensive computer-based
design systems, and the experience gained in
developing expert systems can also be used to
provide a means of incorporating the expert
knowledge of human designers. The challenge
is now to find effective means of combining
such knowledge with that produced by
calculation and simulation to evolve truly
realistic designs.

Computers and Operation

The impact of computers and microelectronics on the instrumentation and control
systems of process plants is even more
evident than their impact on design (see for
example Sargent, 1980). The new hardware
provides transducers for the measurement of
a much wider range of process variables than
hitherto. Incorporation of microprocessors
into instruments allows for direct readings
of quantities based on composite measurements,
and provision of automatic purging, sampling
and recalibration sequences. Solid-state
microelectronics is not only intrinsically
more reliable, but it is cheap to provide
built-in redundancy and self-checking
features.

Digital signals are also more reliably
transmitted, and local intelligence, in the
form of a microprocessor, at each measurement
or control point has brought about a revolution in communication systems. All signals
can be injected into a single high-speed data
highway, and using "packet-switching"
techniques, each "message" has its own
identifier and redundant information for
consistency checking. Each local station can
thus select information relevant to its own
function, and return an acknowledgement of
correct receipt and satisfactory execution of
appropriate action. Apart from its normal
function as a node in a coordinated control
system, each microprocessor can also provide
independent local control in case of
communication breakdown. Fibre-optics will
bring complete freedom from electromagnetic
interference, and a system as safe as the old
pneumatic systems, as well as even higher

performance. Thus we have the capability of building control and communication systems of very high accuracy and integrity, with instant reporting of faults and the ability to institute automatic emergency action if they occur.

Systems and control theory has also made great strides over the last thirty years, its growth rivalling that of the computer world itself, and culminating in the outstanding successes of the Apollo and space-shuttle missions. Again, an excellent review of the state of the art for process control in terms of hardware, software and algorithmic developments is given in the collection of papers in the Proceedings of the Engineering Foundation Conference held at Sea Island in 1981 (Seborg and Edgar, 1982) Only a few general points will be discussed here, and a more detailed consideration of some of the issues will be found in a companion paper (Sargent, 1983b) also presented at the A.I.Ch.E. Jubilee Meeting.

With all the marvellous facilities now at our disposal, it comes as a shock to find that the control algorithm built into all the modern process instrumentation systems is the same algorithm that was in common use fifty years ago, before the age of computers - the familiar "three-term" or PID (proportional-integral-derivative) controller. Indeed, in today's systems the user has much less flexibility and freedom of action in implementing other types of control algorithm than he had with the minicomputer-based control systems of ten years ago! In defence of this, the control engineers in the instrument companies will insist that the PID controller deals perfectly satisfactorily with 80-90% of all process control applications, that most of the control theory developed for aerospace applications is irrelevant to process control, and that many of the advanced control ideas are untried or unproved in real-plant conditions. But how, one may ask, are they ever to be proved in real conditions if the standard hardware precludes their implementation? Are we really convinced that our chemical processes are so utterly different from the navigational and life-support systems of a space-craft that we have nothing at all to learn from putting a man on the moon? Would any self-respecting process plant contractor be content with supplying a plant which had only 80-90% chance of meeting the design specification, without providing any back-up - or at least carefully considering what he would have to

do in case of failure?

Fortunately, there are signs that the next generation of hardware will reintroduce some flexibility at this lowest level of the control system, giving an opportunity for wider use of the more promising regulation algorithms which have been developed in recent years.

The major trend evident at the Sea Island Conference was the shift of emphasis from consideration of individual control problems to consideration of the issues involved in overall plant control. This has certainly been the major thrust of industrial activity in recent years, with the linkage of plant control functions to overall management functions, and two sessions were devoted to this topic at the recent Eurochem 83 Conference. It is generally accepted that a hierarchical structure is appropriate, with substantial computing power at the higher levels to deal with day-to-day production management and longer term planning. At this level too, we meet again many of the trends that have been influencing the organization of design, such as the development of comprehensive structured databases, and a trend towards more realistic predictive models for optimal planning, with corresponding increasing demands for computing power.

Another important function of the control system is the monitoring of plant operation to give warning of abnormal conditions, and monitoring of the condition of both plant and control system to give warning of the development of faults or dangerous conditions. In spite of the improvements in instrumentation, it is obviously not possible to provide direct measurement of everything, and many faults or abnormal situations will have to be inferred from indirect measurements. Such inference usually involves comparing actual measurements with those predicted by an appropriate model, and these models may well be more detailed and comprehensive than those used for control or design.

With today's computing aids it really isn't good enough for the system simply to indicate which measured variables are near or outside "normal operating limits". The system itself should be able to identify many faults and trace the root cause of abnormal situations. In most cases it should then be able to take appropriate action and explain its actions to the operator, only appealing to him for help by sounding the alarm when it cannot identify the cause of the trouble or deal with the situation.

Reliable identification of faults and tracing of root causes is a very difficult task, and quantitative analysis can be greatly aided by qualitative information in the form of fault-trees and event-trees. Here again, as in the design phase, we see scope for the incorporation of expert qualitative knowledge by use of "expert system" techniques. With the use of such aids, it is likely that the situations for which the system appeals to the operator for help will be so complex as to be far beyond the capabilities of the traditional kind of process operator - the Three Mile Island incident illustrates only too well the kind of situation that can arise.

At the Sea Island Conference much attention was given to the question of human factors in process control - how an operator goes about his task, what he looks at, and how the displays can be designed to help him. But in my view there was too much emphasis on how he works, and not enough on what he is required to do or why. With the development of fault detection and diagnosis systems, and further progress of automation to deal with all aspects of plant operation, including start-up and shut-down, the traditional operator's role will disappear. The console will become the plant management's window on plant operation, and the manual tasks about the plant will be maintenance tasks carried out by teams with their own independent displays on plant condition.

Implications for the Profession

This brief survey of design and operation shows that computers have made, and continue to make, dramatic changes in the everyday working life of chemical engineers, and this in turn raises some important questions.

The disappearance of the traditional process operator implies not only a redistribution of tasks between men and computers, but a complete reorganization of plant management. It will not happen overnight, but will evolve through a steady upgrading of the man who sits at the console, who must eventually be a graduate engineer. The question is, what skills will he have to learn that the process operator acquires by experience, and are there quicker and more effective ways of teaching a graduate these skills?

In the world of design, a similar question hangs over the future of the engineering draftsman. In the past he has not only made drawings, but prepared materials take-off and carried out many of the more routine design calculations. Now the computer can do all these tasks, and the man who sits at the work-station will also be a graduate engineer. What skills must he take over if the draftsman is finally to disappear?

In a recent survey of the view of working chemical engineers (Lipowicz and Hughson, 1983), one of the major complaints was that university courses are too theoretical, do not teach enough "practical skills", and do not familiarize students with real plant, equipment and instrumentation. If the above trends are correct, this problem is going to be accentuated, for the graduate engineer in industry will not have the direct support of experienced, practically-minded technicians.

The university has obvious problems in this area, and it is also clear that familiarization with real plant and practical skills can be more rapidly and easily acquired by the new graduate in a properly structured induction programme in his first job. But, although the main responsibility must rest with industry, the university has the responsibility of inculcating the right attitude. If students never see or learn about real equipment, and practical problems are never discussed, the new graduates will either adopt a superior attitude that they are above such matters, or will shy away from dealing with them out of a feeling of inadequacy and inferiority.

To learn state-of-the-art techniques, students need state-of-the-art equipment both in the laboratories and for their computing and design work - and for the latter they also need state-of-the-art software and packages. With the increase in sophistication, and the rapidly accelerating rate of development, obsolescence is a real problem, and the difficulties of funding are obvious enough. But it is not only a matter of funding. The cooperation of industry is required, and (with some shining exceptions) industry could do a great deal more to help the university departments - in providing equipment, releasing software, arranging joint projects involving students, and providing work-experience for students.

In the survey referred to above, most working engineers felt their courses had given them adequate exposure to computers. But closer examination shows that this exposure was at a basic level, and development of computer-based techniques, and computer aids for design and process control, are all regarded as specialist areas for the enthusiasts. As the subject expands and becomes more involved with the technology of computer systems, artificial intelligence, and advanced numerical methods, it is considered more and more to be a fringe activity for chemical engineers. Many of the faculty share this view, and fear that wider use of state-of-the-art software packages in the course will breed a generation of engineers totally dependent on the computer, treating the packages as infallible black boxes, and unaware of the methods embodied in them. Already, they claim, the students are ready to fly to the computer for the least calculation, design projects too easily turn into an orgy of computer programming, computing becomes a substitute for thinking, and students never acquire that intuitive feel for approximations and orders of magnitude so vital to the engineer.

It is an obvious danger, but it need not be like this, and we have to dispel the attitude that engineers do not need to come to terms with the computer, for it is going to dominate their work whatever job they choose. As I have tried to show in the earlier sections, modern developments are making computer packages less like black boxes, with facilities to allow the designer to use his physical intuition, and even his qualitative knowledge. In formulating his models, the engineer must also have some feel for the numerical methods embodied in the packages. It follows that the packages are excellent tools for the development of these skills in the students, and we must teach them enough about the underlying techniques to enable them to use the packages with confidence - and to understand the results.

Intuition is a matter of assimilated experience, and the availability of powerful computer aids makes it possible for today's students to acquire this experience far more rapidly than was hitherto possible. By computer simulation they can explore a variety of cases, examine the effect of changing key design parameters, study the behaviour of control systems - and all this using realistic data and realistic assumptions.

But the issues go much deeper than just providing sophisticated facilities of this sort. We would probably all agree that the key contribution of the chemical engineer is an understanding of the physical situation, and the provision of a suitable model which gives quantitative expression to that understanding. Indeed, the developments in computing have lessened the need for the working engineer to devize calculation methods for himself, and hence served to emphasise this role.

University teachers know only too well how difficult it is to develop in their students the intuition required for effective model-building, and of the woeful tendency of weaker students to try and force their problems into some standard mould, however inappropriate, so as to be able to apply the standard solution methods. We should do better not to teach these stereotyped methods, and use the time instead to develop the students' own abilities.

The students need to learn the conditions under which certain effects can be safely ignored, the dominant mechanisms, and the characteristics of particular structures - and here again the computer can come to our aid. In the laboratory it takes great care to find systems that exhibit the various mechanisms separately, but in a computer simulation mechanisms can be added or dropped at will. For a catalytic reactor, for example, the students can study the dynamics due to the chemical kinetics alone , then add progressively surface transfer, pore diffusion, and finally thermal effects. With a carefully constructed set of such teaching aids, the students could explore for themselves a wide variety of basic physical phenomena and their interactions.

But this must not become a substitute for real experience in the laboratory. On the contrary the use of such aids can make laboratory work a much more fruitful experience. They open up the possibility of studying much more complex systems, even up to trials on pilot plants, using the computer modelling aids to discover just what mechanisms are dominant and how much complexity is required to reproduce the results.

Although I have described these as teaching aids, they are of course precisely the tools the working engineer requires to develop models for real plant, and hence they should also become an essential component of industrial simulation packages.

This is illustrative of an important influence of computers on chemical engineering. They not only enable us to bring more realism into our teaching, but also draw together and illustrate the common needs of teaching, research, and industrial activity. Perhaps this will help to produce graduates better fitted, and better motivated, to work in industry, and reduce industry's suspicions of the utility of academic work.

I do not see computers as a dehumanizing influence, or an enemy of intuition. Rather I see them as a powerful aid to our physical understanding, a new tool whose manifold possibilities we have only just begun to discover. Already we seem to have enough new ideas and new avenues to explore to occupy us until A.I.Ch.E. celebrates its centenary, if not for another 75 years!

REFERENCES

1. Edgar, T.F. and D.E. Seborg (Eds.), (1982) "Chemical Process Control 2", Proceedings of the Engineering Foundation Conference, Sea Island, January 1981. United Engineering Trustees Inc. 1982.

2. Eurochem 83; Chemical Engineering Today - The Challenge of Change. Birmingham, June 1983. I.Chem.E. Symposium Series No. 79. To be published by Pergamon Press

3. FOCAPD-83 - Second Conference on Foundations of Computer-Aided Process Design, Snowmass, June 1983. Proceedings to be published by A.I.Ch.E.

4. Grossmann, I., and M. Morari (1983), "A Dialogue on Resiliency, Flexibility, and Operability - Process Design Objectives for a Changing World", Proceedings, FOCAPD-83 (see above).

5. Lipowicz, M.A. and R.V. Hughson, (1983), "Putting College back on Course - A CE Survey", Chemical Engineering, 90, pp 48-60, September 1983.

6. Sargent, R.W.H., (1980), "Process Control and the Impact of Microelectronics", Chemistry and Industry, 20 December 1980, pp 926-928.

7. Sargent, R.W.H., (1983a), "Process Systems Engineering - Challenges and Constraints in Computer Science and Technology", Proceedings, FOCAPD-83, (see above).

8. Sargent, R.W.H., (1983b), "New Challenges for Process Control", A.I.Ch.E. Diamond Jubilee Meeting, Washington D.C., October-November 1983, Paper no. 61a.

PHOTOCHEMICAL REACTOR ENGINEERING

Joshua S. Dranoff ■

Photochemical Reaction Engineering has been an active field of research and development for at least 30 years. This paper traces some of the major developments in this field and summarizes its current state. Major considerations involved in photoreactor design are discussed and important areas for future development of the field are suggested.

This paper is concerned with the current state of photochemical reactor engineering. This branch of chemical reactor engineering has been an active area of investigation by researchers in this country and abroad for about 30 years and has reached a fairly well developed state. Since most aspects of photochemical reactor design have by now received some attention, someone faced with the task of designing a photoreactor can find guidance in the literature for such an undertaking. The objective of this paper is to discuss the kinds of questions which arise in this context and to point out how they may be pursued. Some perspectives on the development of the field will be presented along with discussion of remaining open problems as well as recent developments and possible new areas of application for photoreactors. However, it should be noted that this is not meant to be a comprehensive review of the literature of applied photochemistry. It is, rather, a personal view of the important developments in the field and its current status.

PHOTOREACTOR ENGINEERING

For present purposes, a photochemical reactor - or photoreactor - is defined as a chemical reactor in which reaction is ini-

Department of Chemical Engineering, Northwestern University, Evanston, Illinois 60201.

tiated by the absorption of electromagnetic radiation in the ultraviolet to visible range (from about 200 to 700 nm) by some molecular species in the reacting medium. Reaction is thus invariably accompanied by attenuation of the initiating radiation; as a result, photoreactors will always exhibit a nonuniform radiation field in the reaction zone despite the presence of physical mixing processes. In this respect, photoreactors differ conceptually from thermal or catalytic reactors in which the presence of nonuniform concentration or temperature fields reflects the efficiency of physical transport processes which are in turn strongly influenced by the motion of the fluid in the reaction zone. Another related special feature of the photoreactor is the requirement of a window through which the initiating radiation can enter the reaction zone. While obviously necessary as the immediate source of radiation, photoreactor windows can also be the seat of problems in cases where opaque deposits form during reaction and eventually bring about reactor shutdown.

In the context of the above definition, photoreactor engineering comprises the selection of reactor configurations and components and the quantitative analysis of reactor performance, given the basic chemistry and reaction kinetics. (In some cases it may also include some efforts at determining kinetics as well.) Presumably the objective of photoreactor engineering is the development of a photoprocess from pilot plant to commercial scale.

It excludes such overall process concerns as determination of feed properties and development of product separation and work-up facilities, but it may well include consideration of heat transfer in the reactor. Certainly involved in photoreactor engineering are consideration of radiation distribution and the efficiency of energy use, analysis of flow and mixing characteristics of the reactor and development of quantitative models for data correlation and design.

INDUSTRIAL APPLICATION OF PHOTOCHEMICAL REACTIONS

While there has been a steady flow of research in this area, particularly in the last 20 years, the instances of industrial application of photochemistry to the production of chemical products, what has been called photochemical synthesis by Fischer [1], have been remarkably few. This obviously excludes such important photochemical applications as printing, photoreduction, and related areas which do not involve photoreactors in the usual sense. Indeed, in view of the scarcity of commercial applications, one might be tempted to view photoreactor engineering as a "set of solutions looking for a problem". However, there has continued to be very extensive research activity in photochemistry and related fields throughout the last few decades, spurred on in part at least by the potential specificity or selectivity of photochemical reactions and the possibility of lower temperature operation than is feasible with thermal or catalytic reaction routes. The high level of activity by photochemists continues to inspire those interested in photoreactors with the promise of potential new applications and the development of commercial processes. Whether such new applications will indeed emerge is of course difficult to predict, but that possibility, as well as the intriguing technical problems associated with photoreactors, have kept engineers working in this field.

With regard to industrial photochemical syntheses, one can mention several which have been described in the open literature. The most prominent applications have been to free radical reactions where the possibility of long chain lengths (high quantum yield) has helped overcome the expense and unconventional nature of the photochemical route. The earliest applications were in photochlorinations. Two examples are the production of the insecticide gamma-hexachlorocyclohexane (also called gamma-benzenehexachloride or BHC) by photo-addition of chlorine to benzene, devel-

oped by the Ethyl Corporation (Governale and Clark [2]) and the Phillips Petroleum process (Hutson and Logan [3]), for production of monochloroalkanes by photosubstitution of chlorine for hydrogen on linear paraffins (in the C11 to C14 range) for later conversion to alkylbenzene sulfonate detergents.

Other free radical photoreactions which have apparently been commercialized include sulfochlorination of paraffins, which is again initiated by chlorine radicals, to yield alkane sulfonyl chlorides and sulfoxidation to produce alkane sulfonic acids, as developed by Hoechst (Fischer [1]).

A non-chain large scale synthetic reaction which has been commercialized is the production of cyclohexane oxime - an intermediate in the production of caprolactam, the monomer of Nylon 6. This photonitrosation involves reaction of nitrosyl chloride with cyclohexane and was first developed by Toray Industries of Japan (Fischer [1]).

Other photoreactions have been commercialized on a smaller scale for the production of pharmaceuticals, fragrances and other specialty chemicals. Perhaps the most famous of these is the synthesis of Vitamin D and related compounds from various steroids (Pape [4], Fischer [1]).

Although the patent literature describes other processes which have been conceptualized and developed to various degrees, there is no discussion of any other large scale commercialized photochemical process in the process engineering literature. While one frequently hears anecdotal comments which suggest more widespread use of photochemical processing steps in the pharmaceutical and specialty chemical fields, there is little or no documentation of these applications. In the absence of published information, one can only note that while such small scale processes may involve unusual reactor designs, they probably do not require a very deep or detailed analysis nor optimization of the reactor. Process economics for the small scale production of high value products may make a detailed and carefully optimized design unwarranted.

REVIEW OF MAJOR DEVELOPMENTS IN PHOTOREACTOR ENGINEERING

Despite the absence of a high level of industrial activity, research and development into photoreactor design and analysis techniques have progressed steadily. Whereas earlier studies were concerned with proper accounting for effects of radiation wavelength on reaction kinetics mainly for homogeneous reacting phases, emphasis in recent years has been on more rigorous modelling of the radiation distribution within the reactor and extension to non-homogeneous systems.

TYPICAL PHOTOREACTOR CONFIGURATIONS

Before discussing these developments further it would be useful to consider briefly some of the photoreactor configurations which have been used in the past for both research and industrial applications. Classification of photoreactor types was first presented by Doede and Walker [5] who pointed out that "optical, chemical and mechanical problems are involved in the design of photoreactors".

The first distinction to be made is that between batch and continuous reactors. Batch reactors clearly find applications only for liquid systems, with semi-batch reactors possible for gas-liquid systems. The batch reactor is typically a stirred vessel irradiated from within or by some external light sources, as illustrated in Figure 1. In the first case one or more light sources may be placed within a protective sheath immersed directly in the reacting medium. The protective sheath thus serves as the radiation window. One variation on this design is the annular batch reactor in which the light source is located along the central axis of a concentric annular reactor vessel. In this case, the inner wall of the reactor serves as the window. Alternatively, the reactor may be irradiated by one or more lamps suitably arranged around the outside of the vessel. In this situation the outer wall of the reactor vessel (usually a circular cylinder) serves as the window.

Mechanical agitation is normally necessary in the batch reactor system. This may be easily provided by the usual centerline mixers in the case of externally irradiated vessels. A more complex mixing arrangement will be needed for an annular design, incorporating multiple mixers or paddle-type stirrers. Yet another way to promote mixing is by rapid circulation of the fluid through an external pump-around circuit.

Such batch reactors can be easily adapted to continuous flow operation by provision of suitably located inlet and outlet ports. They may also be adapted for continuous or semi-batch operation with gas-liquid mixtures. In the latter case, the liquid remains as a fixed batch within the vessel while gas is added continuously. For gas-liquid operation gaseous reactant can be supplied through a suitable sparger and, if present in sufficient quantity, may be used to provide agitation as well, thus eliminating the need for a mechanical mixer, as sketched in Figure 2.

Continuous flow reactors for gas or liquid operation usually take the form of a tubular or annular vessel (See Figure 3). The annular reactor normally has the light source located along its central axis with reactant flowing in the axial direction through the annular space. As in the annular batch reactor, the inner wall of the annulus serves as the radiation window. The tubular reactor is normally a round tube irradiated from the outside. Several light sources may be arranged around the tube for this purpose, or it may be placed within a reflecting enclosure in order to achieve better utilization of the radiation from the lamps. The elliptical photoreactor used first for research studies by Baginski [6] is of this type. This consists of a bright aluminum reflector in the shape of a right elliptical cylinder with the tubular reactor located at one focus of the ellipse and the tubular light source located at the other focus. The rationale for this design is that the light from the lamp which does not impinge directly on the reactor tube surface will reflect off the aluminum surface and then strike the otherwise shaded surfaces of the tube. This enhances the utilization of the light emitted by the lamp and produces a more uniform irradiation of the reactor than would otherwise be possible with a single light source. While other types of flow reactor involving thin liquid films, parallel plate channels, etc., have been proposed and used, those mentioned above have received the most attention.

The previously mentioned BHC process developed by the Ethyl Corporation used the annular reactor configuration. Fluorescent lamps provided the radiant energy for this system. They were located in a 2 inch diameter Pyrex tube surrounded by a 4 inch diameter Karbate tube through which the reactant mixture of chlorine and benzene flowed. An 8 inch diameter steel tube completed the concentric annular arrangement and carried cooling water for temperature control. The reactor tubes were arranged in a continuous folded stack to provide the necessary reaction volume.

A variant of the elliptical reflector configuration was used in the Phillips Petroleum alkane chlorination process. It involved a single lamp at the center and 4 parallel vycor reactor tubes located around it within a bright circular reflector.

Development of Photoreactor Literature

With this background, let us know consider briefly some of the important developments which have been reported in the photoreactor engineering literature. A noteworthy paper which in some sense marked the onset of relatively widespread research in this field was the work of Huff and Walker [7] on the photochlorination of chloroform. They investigated this vapor phase reaction in an elliptical photoreactor of the type described above, thus helping to bring this type of device to the attention of other workers. (It should be noted that Gaertner and Kent [8] used an elliptical reflector but with an annular reactor instead of a circular tube a few years earlier.) Huff and Walker dealt in a quantitative way with the complex kinetics of the free radical chain reaction and with a polychromatic light source. They showed how wave length distributions (lamp output spectra) can be taken into account and demonstrated the use of quantitative actinometry to characterize the reactor system. They did, however, make the simplifying assumption that radiation from the lamp was emitted radially in planes perpendicular to the lamp axis, with the result that radially incident uniform light was assumed to illuminate the reactor. Similar assumptions were subsequently used by a number of other investigators, despite acknowledged shortcomings, in the absence of more appropriate models.

Studies of annular batch reactors for liquid systems were initiated in our laboratories by Harris [9] and Jacob [10,11,12] who studied the photolysis of chloroplatinnic acid and later by Jain [13] and Daniil [14] who investigated photopolymerization in that geometry. The work of Jacob [12,15] marked the beginning of efforts to model lamp output in a more fundamental manner. An isotropic line source was used in place of the radially directed lamp model with improved results.

The importance of mixing and diffusional effects in photoreactor design were the subjects of a number of theoretical papers by Hill and coworkers, following the inital work of Hill and Felder [16]. These papers [17,18,19,20] introduced the concepts of macro and micromixing in the photoreactor context and showed under what circumstances mixing and diffusional effects are of importance.

At about the same time, J.M. Smith and co-

workers began an extensive series of important publications on photoreactor engineering with a study by Cassano and Smith [21] of the vapor phase photochlorination of propane in an elliptical photoreactor. Smith and coworkers continued the study of photochlorinations (Cassano [22], Boval [23], Santarelli [24]) and also studied photodecomposition reactions aimed at destruction of undesirable organic solutes in waste waters (Matsuura [25,26], Schorr [27], Boval [28]). Through use of multiple annular tube arrangements they showed how the wavelength distribution of the light which ultimately reaches the reactor windows could be controlled by means of filter solutions of various types. They also explored in detail the methods for accounting for wavelength dependent kinetics and radiation attenuation characteristics, and investigated the effects of wall deposits on reactor performance (Ziolkowski [29]). Cassano, Sylveston and Smith [30] also authored an important paper which reviewed all of the work done prior to 1967, outlined the essential features of photoreactor design and pointed out some areas requiring further attention. In another paper of some utility, Matsuura and Smith [31] considered the effect of assumptions concerning the radiation model on the analysis of kinetic data. Based on analysis of two different models, they suggested that conclusions concerning reaction kinetics will not be sensitive to the radiation model as long as the same model is used for analysis of actinometric data as well as kinetic data for a given reaction system.

Measurement of Radiation Fields. At about this time many researchers began to investigate more thoroughly the adequacy of the rather simplifed models which had been used to describe radiation distribution profiles. This effort was sparked, in part at least, by the first reported efforts to measure such distributions by Jacob and Dranoff [12,15]. Using a small selenium barrier photocell Jacob [12] made measurements which reflected the radiation profiles in an annular reactor vessel used for batch liquid reactions and showed that predictions based on a line source model for the centrally located lamp were somewhat in error due to neglect of reflection and refraction effects. After empirical correction, these predictions could be used for accurate analysis of the reactor performance. Jacob [15] also made measurements in an elliptical reactor which demonstrated significant errors in the usually assumed radically incident light model for that geometry. His results were, of course, specific to the particular reactor system he studied; small

changes in system geometry due to use of different materials or methods of construction
might well change the results significantly,
although generally similar effects would be
expected.

Those studies were followed by actinometric measurements made by Williams and co-
workers (Zolner [32]) in which the angular
nonuniformity of incident light in an elliptical photoreactor was measured by masking different sections of the reactor tube; more
fine grained measurements within the tube were
also made using a small diameter (1 mm) quartz
actinometer tube connected to an externally
recirculated reservoir of actinometric solution
(Williams [33]). These measurements again
showed nonuniformities associated with the
elliptical reactor in both angular and longi-
tudinal directions.

Akehata and Shirai [34] pointed out the
need to differentiate between specular and
diffuse light sources, particularly when
dealing with fluorescent lamps which are best
represented as a diffuse source.

Most recently Tournier et al. [36] published a paper detailing their development and
use of a novel radiation probe involving a
very small (0.5 mm) diffusing sphere attached
to an optical fiber which transmits its output
to a photomultiplier tube for signal amplifi-
cation. They used this type of measuring
device to map the radiation field in a stirred
reactor and showed that an empirical correction was necessary to obtain agreement between
measured data and simple model predictions.
This type of device, with a solid viewing
angle of almost 4 pi, may offer the best probe
for making such measurements with minimum
correction.

Modelling of Radiation Fields. During
this period more detailed and appropriate
analyses of the radiation field in photoreactors were published by two former students of
Smith who have continued their own work in
this field. In a very significant paper,
Cassano et al. (Irazoqui et al. [37]) first
proposed the three dimensional, or so-called
extense source, model for radiation sources.
They pointed out how more rigorous analysis
taking into account both the direction of the
individual ray or ray bundle and the area of
the radiation receiver could lead to improved
prediction of radiation distributions. They
formulated models for several geometries including the annular reactor (Irazoqui et al.
[38]) and later the elliptical reactor (Cerda
et al. [39]) taking into account first reflec-

tions and carried out extensive numerical computations via ray tracing methods to verify
that this improved the prediction of the
distributions compared to simpler models.

Subsequently, Santarelli and coworkers
(Bandini et al. [40]) presented a somewhat
more formal analysis of the photoreactor problem and demonstrated the interaction of the
radiation distribution model and the resultant
concentration field. They also carried out
extensive computations to demonstrate the
application of their model to reactors of
various geometries including in some cases the
effect of light scattering in non-homogeneous
media.

In more recent studies, Gebhard [41] has
shown how the radiation fields from real fluorescent and germicidal lamps must be corrected
to account for nonidealities in lamp construction and operation and how reflection and
refraction effects can be taken into account
through ray tracing techniques, again using
the digital computer to carry out the tedious
numerical computations. His work was extended
by Pons [42] who refined the numerical techniques involved and also considered problems
associated with a multiple lamp configuration.

Newer Developments. While the above
investigations have concentrated on various
aspects of essentially traditional photoreactor design, three other areas which have been
studied in recent years may have a major
impact on photoreactor development. Almost
all the work now in the literature has been
concerned with the behavior of single phase
systems. However, many applications, such as
photochlorination for example, involve the
contact of gas and liquid phases in the
presence of light. Detailed analysis of such
systems has been quite limited until recently.
An important contribution in this area was
made by Akehata et al. [43] who studied light
transmission through aerated and packed photoreactors and derived a model to account for
scattering phenomena. More recently, Yokota
et al. [44] considered the bubble photoreactor
in greater detail and showed how the radiation
profiles in such devices may be obtained by
use of a computerized Monte Carlo technique.
They applied their method of analysis to analyze the behavior of a toluene chlorination
reactor with good results. Santarelli et al.
(Spadoni et al. [45]) have also done theoretical analysis of light scattering using Monte
Carlo techniques. Because of the importance
of multiphase systems this area merits further
attention, particularly with respect to comparisons between experimental measurements and

model calculations.

A second area of increasing interest is the field of photocatalysis. Considerable effort is currently being expended by many researchers seeking ways to capture and store solar radiation by producing chemical changes. Much of this work centers on identification of the proper reaction chemistry and catalysts to promote desired reactions and is beyond the scope of this paper. However, some investigators have considered the behavior of photocatalytic reactors and the difficult design question which they may pose. Hacker and Butt [46] made an initial study of a slurry type reactor involving the reduction of methylene blue on ZnO particles and demonstrated that the optical characteristics of the slurry could be described by models for homogeneous phases by proper choice of an attentuation coefficient. A recent paper by Daroux et al. [47] presented a new reactor for solid catalyzed vapor phase reactions involving a dilute moving bed of catalyst stirred by a special agitator. This unit was designed to permit maximum irradiation of the solid-gas mixture.

More recently Ollis and coworkers have explored two different types of photocatalytic reactors. In one set of experiments, Pruden and Ollis [48] have shown that slurry reactors containing TiO_2 can serve as excellent means for oxidation of organic compounds in dilute solutions. They have used an externally irradiated annular reactor for this work, which reopens the possibility of using photoreactors for waste water treatment. Coincidentally, there also has been a surge of interest in the use of photoreactors for disinfection and sterilization of waters - an area of interest to both chemical and environmental engineers (Sugawara et al. [49,50], Severin [51]).

Ollis and Marinangeli [52,53,54] have also proposed and analyzed an innovative type of reactor for photocatalyzed rections which involves a catalytically active surface coating applied to the outside of an optical fiber. Their design called for transmission of initiating radiation (possibly solar in origin) along an optical fiber with some penetration of the light ray from within the fiber into the coating at each reflection - sufficient to catalyze reaction between the surface and a surrounding fluid phase. No practical application of this configuration has emerged as yet but it appears to be an ingenious device for overcoming the inherent opacity of catalyst slurries and may prove to be quite useful.

Finally, a new approach to reactors for free radical initiated reactions which are frequently subject to problems of deposit and by-product formation was proposed by Lucas [55]. He suggested that photoreactors be divided into two distinct zones: a photoinitiation zone where radicals or other active species are formed by interaction of radiation and the appropriate reactant species, and a thermal reaction zone, shielded from radiation, to which the active species would be transported to initiate subsequent reaction. In his early experiments and proposals the separation was made by a Teflon diaphragm. He applied this idea to the photooximation reaction described earlier and reported successful operation of that gas-liquid system and elimination of the problem of deposit formation on the radiation window. We (Dworkin and Dranoff [56], Mazich [57], and Chakravaty [58]) and one other laboratory in France (Richard [59]) have subsequently investigated application of this technique to gas phase reactions - in particular, the chlorination of methane and chloroform - and have also found that it may offer promise. Work on this approach continues. It has suggested to us that similar results may be achieved by control of flow patterns within the photoreactor. The utility of the two zone reactor approach and indeed the design of a more general free radical generator for use in other contexts has been discussed further by Lucas [60,61]. Its ultimate area of application remains to be determined in the future, however.

Yet another variation on "traditional" photoreactor design may result from current studies of laser initiated chemical transformations. The existence of coherent laser prodcued radiation has had a very significant impact on photochemical research and it is likely that this technology will have strong influence on photoreactor design as well. While researchers have begun to consider such problems, no detailed studies of an engineering nature have as yet been published. Without doubt, however, more will be heard from this field in the future.

STEPS IN PHOTOREACTOR DESIGN

The picture which emerges from this selective summary of developments in photoreactor engineering is one of a field which has progressed steadily in the last 25 years to the consideration of more accurate models of the radiation field and the more demanding multiphase problems. As researchers have looked more deeply at photoreactors, they have sharpened their analyses and have demonstrated how simplifying assumptions can be eliminated in order to make reactor models more accurate and useful. In view of this situation, it is fair to state that good tools are now available for the analysis and design of photoreactors which can provide guidance to the design engineer. To emphasize this point of view, let us consider the steps required in the design of a photoreactor for commercial purposes.

Naturally such efforts must begin with the identification of a photochemical reaction which is to be developed further. Presumably the chemistry involved will have been investigated to the point where the reaction stoichiometry will be known along with information on the effective radiation wavelengths, the physical properties of reactants and products including those which affect transmission and absorption of radiant energy, and the presence or absence of photosensitizing species. In other words, reaction conditions and anticipated product distributions should be known as a result of basic chemical investigations.

Also, the kinetics of the reaction should be known if design is to proceed. This infers the existence of a reaction rate model which accounts for the effects of reactant and product concentrations and the role, if any, of mass transfer phenomena. For many photoreactions the rate expression takes the following simplified form:

$$r = k (Ia)^m C^n$$

where Ia, frequently called the absorbed intensity, has been traditionally used to represent the volumetric rate of absorption of radiation by the reacting medium. The parameters of the rate model in general will be wavelength dependent; presumably such information will also be known.

In some cases determination of the reaction kinetics may not as yet have been completed and this may become one of the objectives of a pilot-plant program. Due caution should be exercised in that situation in the design of the pilot reactor system. Proper elucidation of the reaction kinetics demands a reactor system in which fluid phase compositions and rate of energy absorption can be unequivocally determined. This suggests the use of the most simple geometry possible for the photoreactor, such as a one dimensional (parallel plate) reactor vessel illuminated by plane parallel radiation. In the event that this type of reactor is not available for the experiments at hand and a prototype reactor must be used, it will be necessary to resort to complex modelling in order to fit the potential kinetic models to the data. This situation is, of course, exactly analogous to what might be encountered with a thermal reaction. Kinetic data are best obtained under isothermal conditions in nonflow reactors with uniform compositions. If a flow reactor is used instead, then uncertainties in flow distribution, residence times, temperature history, and diffusional effects may all combine to make kinetic analysis difficult or at least less clearcut. Hence, if kinetic data are needed, experiments should be carried out in equipment expressly designed for that purpose if at all possible.

Once the reaction chemistry and kinetics are known, the work of the photoreactor engineer normally begins. The first tasks are to select the lamps to be used and to decide upon the proper reactor configuration. Lamp selection will be determined by the radiation wavelength range required and the anticipated production rate for the reactor. The former is obviously critical for initiation of the desired reaction as well as avoidance (if possible) of undesired side reactions or product degradation steps. The latter will affect directly the required power output of the lamp. In turn, a decision on the lamp power will determine the need to cool the lamp to maintain its proper operating temperature.

The determination of reactor configuration is perhaps most critical and will depend on factors such as:

1. the choice of batch or continuous operation and required throughout
2. the nature of the phases involved in the reaction
3. the necessity for heat transfer (removal or addition) to reactor contents to control temperature
4. the necessity for heat removal from the lamp
5. the necessity for mixing of the reactor contents to eliminate composition and temperature gradients
6. the light absorption characteristics of the reacting medium
7. the chemical nature of the reactants and products and their corrosivity
8. the normal temperature and pressure of the system
9. the likelihood of window deposit and byproduct formation.

When these factors are known or stipulated, a decision as to reactor type can be made. One of the standard photoreactor types

described earlier may be selected or an alternative, perhaps novel, configuration will be
decided upon in order to accommodate these
physical and chemical factors.

As an example, suppose the reaction in
question is a vapor phase chlorination of a
hydrocarbon-like molecule taking place at a
temperature of about 200°C and a pressure of 1
atm. If the reaction is only mildly exothermic, it might well be carried out in a tubular
flow reactor made of Pyrex glass and irradiated from the outside. A "black light" near-
ultraviolet light source could be used to
activate chlorine and promote this reaction
since the radiation at 350 nm given off by
such a lamp is known to be effective for
chlorine activation. Perhaps an array of parallel reactor tubes and fluorescent lamps
placed reasonably close together in a regular
pattern and surrounded by a reflecting shield
would be suitable. Air could easily be circu-
lated across the tube bundles to cool the
lamps and remove the heat of reaction. Pyrex
pipe would be sufficiently strong at reaction
conditions and resistant to chlorine containing vapors and would also pass most of the
effective radiation with minimum absorption.
Decisions on lamp placement and tube size
would require further analysis.

However, should the reaction be highly
exothermic, heat removal from the reactor
would undoubtedly be necessary. In that case
a better reactor might be an externally irradiated annular type with cooling fluid circu-
lated through the inner tube where it would
not interfere with radiation transmission.

Or, suppose the process involves the reaction of a gas such as oxygen with a second
liquid phase reactant. In this case, a two
phase reactor must be utilized in which the
gaseous reactant is dispersed in the form of
fine bubbles in the liquid by a suitable
sparger-agitator system irradiated by several
lamps placed in vertical light wells position-
ed appropriately - with due regard for the
presence of agitators as well as the light
attenuation characteristics of the reactant
medium.

Clearly, these suggestions only illustrate
the role played by some of the factors mentioned earlier in the selection of reactor
configuration. This is obviously a step in
the design where the ingenuity of the engineer
is most critical.

Once reactor configuration is fixed,
attention may be turned to quantitative analysis of the anticipated reactor performance.
This step is ultimately aimed at determination
of the reactor size and product throughput -
as with any other type of chemical reactor.
The considerations involved at this point are
to a large extent the same as for thermal or
catalytic reactors with the added complication
of determination of the radiation distribution. That is, it is necessary to determine
the distributions of fluid velocities, temperatures, and concentrations as well as radiation density in order to complete a reactor
analysis. The fluid mechanical and heat
transfer aspects of the reactor will involve
no new or different principles. Indeed, even
the concentration profile equation is essentially no different in this case with the
exception of the form of the reaction term.
It is only in the specification of the radiant
energy distribution equation that we encounter
a distinctly different equation.

Following the analysis of Santarelli et
al. (Bandini et al. [40]), one may write all
of the equations which describe a photoreactor, including those which represent the
radiation profile. These equations turn out
to be strongly coupled in general because of
the interdependence of the concentration and
radiation profile relations. Under some special circumstances, such as when a photosensi-
tizer is the only absorbing species present,
the equations uncouple and solution is not
difficult. Otherwise, however, solution will
require the use of digital computer techniques, involving ray tracing methods in general. Details, including methods for handling
the cases of scattering, refraction, and
reflection, are best left to the original literature at this point. In many cases a simplified analysis of the radiation distribution
may be adequate for preliminary design purposes. Parallel plane radiation models, although obviously not correct, may nonetheless
serve as adequate preliminary models for estimating reactor performance. However, when
quantitative designs are needed, more rigorous
models should be used, particularly in view of
the widespread availability today of the
required computing resources.

FUTURE WORK ON PHOTOREACTOR ENGINEERING

As suggested earlier, some aspects of
photoreactor engineering which are likely to
receive increased attention in the future are
photocatalysis, for both chemical synthesis
and water purification, the use of lasers in
photoreactors, and the development of photo-
biological reactors for sterilization of waste
and drinking waters. These hold promise for

greater application of this still relatively unfamiliar technology. Continued use of photoprocesses for production of specialty materials including pharmaceuticals is likely to remain on a small scale and thus will probably not inspire any large scale research effort in photoreactor engineering. Finally, work on novel photoreactor configurations and the effects of flow patterns on reactor performance will undoubtedly be pursued to a small extent in the future as well.

But it seems most likely that any large scale application will await the discovery of practical laser induced reactions; development of such reactions could well lead to more intense research on the engineering of photoreactors. Otherwise this technology will continue to exist for a variety of relatively small special purpose applications which will develop at a steady but slow rate, with research and the literature of photoreactor engineering developing at a corresponding pace.

LITERATURE CITED

1. Fischer, Martin, Angew. Chem. Int. Ed. Engl., 17, 16 (1978).

2. Governale, L.G. and J.T. Clark, Chem. Eng. Prog., 52, 281 (1956).

3. Hutson, Thomas and R.S. Logan, Chem. Eng. Prog., 68, 76 (1972).

4. Pape, Martin, Pure Appl. Chem., 41, 535 (1975).

5. Doede, C.M., and C.A. Walker, Chem. Eng., 62, 159 (1955).

6. Baginski, F.C., D. Eng. Dissertation, Yale University, New Haven, Connecticut (1951).

7. Huff, J.E. and C.A. Walker, AIChE J., 8, 193 (1962).

8. Gaertner, R.F. and J.A. Kent, Ind. Eng. Chem., 50, 1125 (1958).

9. Harris, P.R. and J.S. Dranoff, AIChE J., 11, 497 (1965).

10. Jacob, S.M. and J.S. Dranoff, Chem. Eng. Prog. Symp. Ser., 62, 47 (1966).

11. Jacob, S.M. and J.S. Dranoff, Chem. Eng. Prog. Symp. Ser., 64, 54 (1968).

12. Jacob, S.M. and J.S. Dranoff, AIChE J., 16, 359 (1970).

13. Jain, R.L., W.W. Graessley and J.S. Dranoff, Ind. Eng. Chem. Prod. Res. Develop., 10, 293 (1971).

14. Daniil, Apostolos, W.W. Graessley and J.S. Dranoff, Photopolymerization of Vinyl Acetate in a Well-Stirred Reactor", Chem. Reaction Engineering, B7-49, Elsevier Publishing Co., Amesterdam (1972).

15. Jacob, S.M. and J.S. Dranoff, AIChE J., 15, 141, (1969).

16. Hill, F.B. and R.M. Felder, AIChE J., 11, 873 (1965).

17. Hill, F.B. and N. Reiss, Can. J. Chem. Eng., 46, 124 (1968).

18. Hill, F.B. and N. Reiss, AIChE J., 14, 798 (1968).

19. Hill, F.B. and R.M. Felder, Chem. Eng. Sci., 24, 385 (1969).

20. Hill, F.B. and R.M. Felder, Ind. Eng. Chem., 9, 360 (1970).

21. Cassano, A.E. and J.M. Smith, AIChE J., 12, 1124 (1966).

22. Cassano, A.E. and J.M. Smith, AIChE J., 13, 915 (1967).

23. Boval, Bruno and J.M. Smith, AIChE J., 16, 553, (1970).

24. Santarelli, Francesco and J.M. Smith, Chem. Eng. Commun., 1, 297 (1974).

25. Matsuura, Takeshi and J.M. Smith, Ind. Eng. Chem. Fundam., 9, 252 (1970).

26. Matsuura, Takeshi and J.M. Smith, AIChE J., 16, 1064 (1970).

27. Schorr, Victor, Bruno Boval, Vladislav Hancil and J.M. Smith, Ind. Eng. Chem. Proc. Des. Develop., 10, 509 (1971).

28. Boval, Bruno and J.M. Smith, Chem. Eng. Sci., 28, 1661 (1973).

29. Ziolkowski, Dariusz, A.E. Cassano and J.M. Smith, AIChE J., 13, 1025 (1967).

30. Cassano, A.E., P.L. Sylveston and J.M. Smith, Ind. Eng. Chem., 59, 18 (1967).

31. Matsuura, Takeshi and J.M. Smith, AIChE J., 16, 521 (1970).

32. Zolner, W.J. and J.A. Williams, AIChE J., 18, 1189 (1972).

33. Williams, J.A., AIChE J., 24, 335 (1978).

34. Akehata, Takashi. and Takashi Shirai, J. Chem. Eng. of Japan, 5, 385 (1972).

35. Funayama, Hitoshi, Kojiro Ogiwara, Takuo Sugawara and Hiroyasu Ohashi, Kagaku Kogaku Ronbushu, 3, 354 (1977).

36. Tournier, A., X. Deglise, J.C. Andre and M. Niclause, AIChE J., 28, 156 (1982).

37. Irazoqui, H.A., Jaime Cerda and A.E. Cassano, Chem. Eng. J., 11, 27 (1976).

38. Irazoqui, H.A., Jaime Cerda and A.E. Cassano, AIChE J., 19, 460 (1973).

39. Cerda, Jaime, H.A. Irazoqui and A.E. Cassano, AIChE J., 19, 963 (1973).

40. Bandini, E., C. Stramigioli and Francesco Santarelli, Chem. Eng. Sci., 32, 89 (1977).

41. Gebhard, T.J., Ph.D. Dissertation, Northwestern University, Evanston IL (1978).

42. Pons, Marie-Noelle, M.S. Thesis, Northwestern University, Evanston IL (1980).

43. Akehata, Takashi, K. Ito and A. Inokawa, "Analysis of Aerated and Packed Photo-chemical Reactors", AIChE Annual Meeting, Washington DC, December 1974.

44. Yokota, Toshiyuki, Toshihiko Iwano, Hiromi Deguchi and Teiriki Tadaki, Kagaku Kogaku Ronbushu, 7, 157 (1981).

45. Spadoni, G., E. Bandini and Francesco Santarelli, Chem. Eng. Sci., 33, 517 (1978).

46. Hacker, D.S. and J.B. Butt, Chem. Eng. Sci., 30, 1149 (1975).

47. Daroux, M., Y. Parent and D. Klvana, Chem. Eng. Commun., 501 (1980).

48. Pruden, A.L. and D.F. Ollis, J. of Catal., 82, 404 (1983).

49. Sugawara, Takuo, Michio Yoneya, Hiroyasu Ohashi and Shigenori Tanagawa, J. Chem. Eng. Japan, 14, 400 (1981).

50. Sugawara, Takuo, Michio Yoneya and Hiroyasu Ohashi, J. Chem. Eng. Japan, 14, 406 (1981).

51. Severin, B.F., Ph.D. Dissertation, University of Illinois, Champaign-Urbana, Illinois (1982).

52. Marinangeli, R.E. and D.F. Ollis, AIChE J., 23, 415 (1977).

53. Marinangeli, R.E. and D.F. Ollis, AIChE J., 26, 1000 (1980).

54. Marinangeli, R.E. and D.F. Ollis, AIChE J., 28, 945 (1982).

55. Lucas, Georges, Informations Chimie, 16, 33 (1971).

56. Dworkin, Daniel. and J.S. Dranoff, AIChE J., 24, 1134 (1978).

57. Mazich, K.A., M.S. Thesis, Northwestern University, Evanston, Illinois (1980).

58. Chakravarty, Dipak, M.S. Thesis, Northwestern University, Evanston, Illinois (1981).

59. Richard, M., Doct. Ingenieur Thesis, L'Institut National polytechnique de Toulose, Toulose, France (1982).

60. Lucas, Georges, Information Chimie, No. 126, 87 (1975).

61. Lucas, Georges, "Transfer Theory for Trapped Electromagnetic Energy", Technique et Documentation (Lavoisier), Paris, France (1982).

BATCH - LIQUID

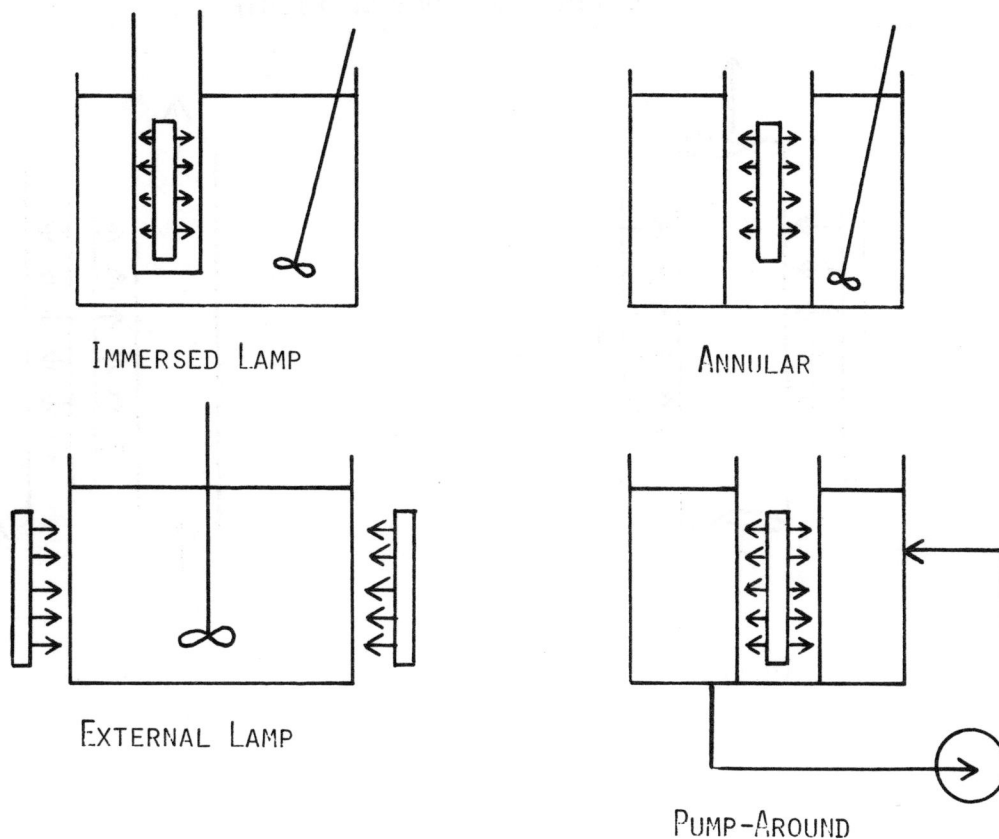

Figure 1. Batch Reactor Configurations.

SEMI-BATCH,   CONTINUOUS

GAS-LIQUID   LIQUID

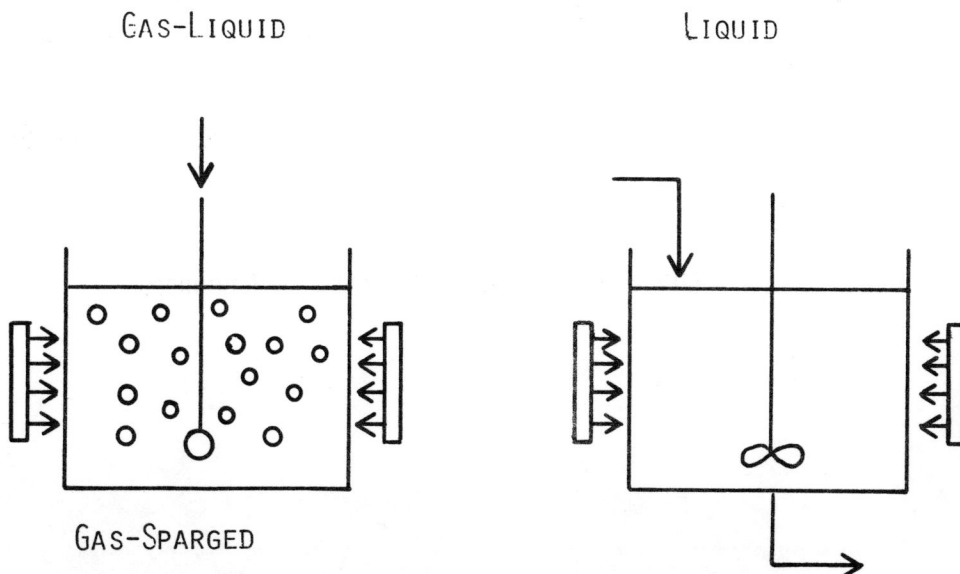

Figure 2. Semi-Batch and Continuous Reactor Configurations.

CONTINUOUS - GAS OR LIQUID

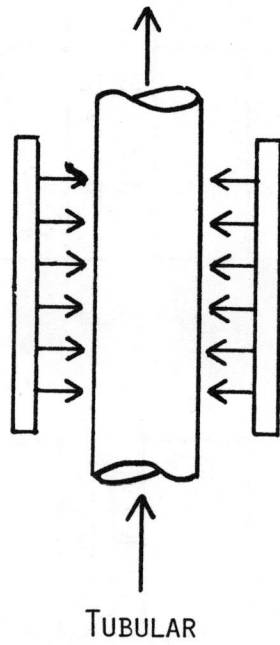

TUBULAR

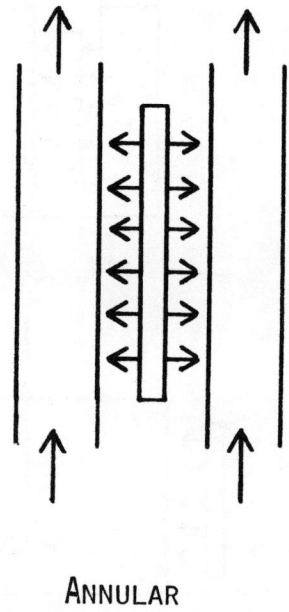

ANNULAR

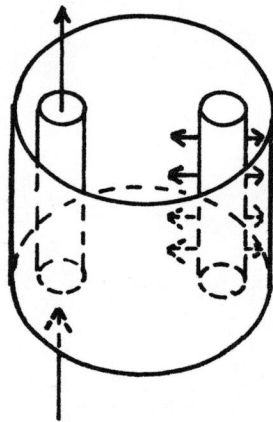

TUBULAR REFLECTOR

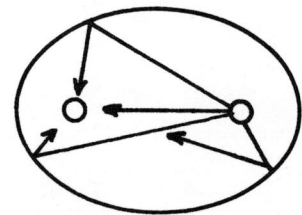

ELLIPTICAL REFLECTOR

Figure 3. Continuous Batch Reactor Configurations.

PERSPECTIVES ON CRYSTALLIZATION IN CHEMICAL PROCESS TECHNOLOGY

Hugh M. Hulburt ■ Department of Chemical Engineering, Northwestern University, Evanston, IL 60201

In the past five decades, the scientific study of transitions to and from the solid phase has grown from an esoteric speciality of a few physicists to a wide-ranging body of theory and experiment with active applications in many engineering disciplines. The present review will focus on just one of these: crystallization of solids from solution. It is one of the oldest of the alchemists arts, one of the most ubiquitous of the unit operations, and one of the latest to begin to acquire a solid scientific foundation for developing new technology. This paper will trace the stages of this development and cast a forward glance at some of the prospects for the future evolution of this historic technical art.

CRYSTAL GROWTH KINETICS

Scientific interest in phase changes may be said to have begun its modern phase in the early years of this century. Even earlier, Gibbs (1878) presented the thermodynamics of growing crystal surfaces under equilibrium conditions. However, he was well aware of the limitations of thermodynamics in describing crystal habits, as the following quotation indicates:

"On the whole, it seems not improbable that the form of very minute crystals in equilibrium with solvents is principally determined by equation (665), (i.e., by the condition that crystal except so far as the case be modified by gravity or solvent no more supersaturated than is necessary to make them grow at all), the deposition of new matter on the different surfaces will be determined more by the nature (orientation) of the surfaces and less by their size and relations to the surrounding surfaces ... The kinds of surface thus determined will probably generally be those for which (the surface free energy) has the least values. But the relative development of the different kinds of sides, even if unmodified by gravity or the contact of other bodies, will not be such as to make the (total surface free energy) a minimum. The growth of the crystal will finally be confined to sides of a single kind.

It does not appear that any part of the operation of removing a layer of molecules presents any special difficulty so marked as that of commencing a new layer; yet the values of (liquid phase chemical potential of the solute) which will just allow the different stages of the process to go on must be slightly different, and therefore, for the continued dissolving of the crystal value of (the liquid phase chemical potential of the solute) must be less (by a finite quantity) than that given by equation (665) (the equilibrium value). It seems probable that this would be especially true of those sides for which the (surface free energies) have the least values. The effect of dissolving a crystal (even when it is done as slowly as possible) is there-fore to produce a form which probably differs from that of theoretical equi-librium in a direction opposite to that of a growing crystal."

These prophetic words are more remarkable when we realize that the atomic or molecular structure of crys-tals was yet to be demonstrated by von Laue 34 years later with the discovery of X-ray diffraction. The conse-quences of Gibbs' criteria, worked out by Curie (1885), and Wulff (1901) made it clear that real crystals almost never display the equilibrium habit. However, it remained for Volmer (1934, 1939), Kossel (1927, 1928), Stranski and co-workers (1928, 1949), Becker and Doring (1935), and Frenkel (1945, 1946) to develop the atomic theory of crystal growth on a more realistic basis. All of these were concerned primarily with the growth of perfect crystals and the theoretical calculation of the surface free ener-gies of partially grown surfaces of differing atom densities, on the as-sumption (dating to Bravais (1866)) that the planes of greatest density will also be those of greatest surface free energy, and hence of slowest growth and dominant in the stable crystal habit. (For a concise sum-mary of these theories, see Verma (1953), Ohara and Reid (1973), and Jackson (1979) and, for recent work Bennema (1983).)

These perfect crystal theories all suffered from the practical defect that they predicted the slow step in the growth process to be the nuclea-tion of a stable cluster on the sur-face, of such size that further growth would no longer increase the specific surface free energy of the crystallite. To form such a cluster at an observa-ble rate by statistical fluctuations in the deposition process was found

to require a fairly large super-saturation. Yet many crystals are known to grow rapidly at almost un-detectable supersaturation. It was soon realized that, while there is only one way to form a perfect crystal structure, there are Avogadro's number of ways to incorporate an imperfection in the lattice. Hence, on statistical grounds, almost the only experimentally observed crystals should be imperfect. The role of imperfections in the growth process should be very impor-tant. This led to the recognition that a dislocation, in which a crystal lattice is displaced slightly by sliding a portion along a line per-pendicular to a growing face, produces a step originating at the dislocation axis which can move with a spiral mo-tion as atoms deposit in it, but does not disappear. Furthermore, the added free energy for deposition along the step is much less than for depo-sition in the middle of the plane surface. Hence growth should be much faster than when a stable cluster must form at random. These concepts were formalized in the dislocation theory of crystal growth by Burton, Cabrera and Frank (1951). An abun-dance of direct observation of the growth spirals has given strong con-firmation of the BCF theory in numer-ous instances for growth on macro-scopic crystal surfaces. Some ques-tions of detail remain in the theoretical predictions (Bennema et al., 1976) but there is little doubt that dislocations play a central role in many crystal growth processes.

On the global scale, however, it is easy to imagine circumstances in which the atomic deposition processes on the surface may not be the limiting step in the overall deposition rate. Long before the atomic processes were seriously discussed, Noyes and Whitney (1897) and Nernst (1904) proposed that crystal growth and dissolution were limited by diffusion processes in the fluid by which the solute was trans-ported to the phase interface. If the incorporation processes on the crystallite surface are much faster than the transport of solute to the interface, then one might assume that the concentration at the interface is essentially that of equilibrium and the driving force for transport is the supersaturation of the bulk solution. The deposition rate per unit area is thus linear in the supersaturation and the deposit thickness on a plane will **grow** at a constant rate. This simple macroscopic model was found to be obeyed, at least approximately, by many systems and has been widely used by physical chemists and biologists. In chemical engineering, it has been enshrined as McCabe's Law (McCabe, 1929; Randolph and Larson, 1971). However, as a macroscopic law it is honored as often in the breach as in observance. While it serves as a good starting assumption in the absence of information, it ought to be checked experimentally whenever its failure might be critical for design or operation.

From these points of view, crystal growth is an example of a molecular interfacial process, involving the usual steps of transport to the surface, migration, and molecular interaction on the surface. In the simplest cases, one or the other of these steps will be rate limiting, and the overall growth rate will show a simple kinetic form. However, in crystal growth it is more likely that two or more steps will have comparable rates and the kinetics will assume a complex form, perhaps approximated by a power law with a non-integer apparent order with respect to supersaturation. The theoretical synthesis of transport processes near and on the surface with the surface molecular processes of growth has been most fully treated by Bennema (1967a, 1967b, 1968, 1969) and by Gilmer et al. (1971, 1972). The dependence of the crystal growth rate on the crystal size, a complex question of which we shall have more to say later, results from the effects of size on the rate of each of the steps in the growth process. It is probably the fortuitous result of cancelling effects that the global growth rate appears to be independent of size in many cases.

NUCLEATION

While Gibbs' principle of minimum surface free energy appears not to be controlling for the macroscopic crystal habit, it survives in full force in our understanding of the rate of initiation of new crystallites in supersaturated solutions or melts. The basic ideas in the theories of nucleation all stem from the circumstance that small aggregates have higher free energies than single particles, but lower free energies as well as higher than large aggregates.

There is thus a free energy barrier to overcome in order to produce a growing crystallite. Aggregates on the small size of this range are known as embryos; those on the large size are known as crystallites; and those on the boundary, for which the free energy change of adding an additional particle is zero, are known as "critical embryos" or "critical nuclei". All "zero-order" theories for the rate of this nucleation process begin with the assumption that the free energy barrier is high enough that the energy distribution of sub-critical embryos is substantially that of an equilibrium system and can be estimated by thermodynamic or statistical mechanical theories. Thus the concentration of critical nuclei can be determined as a function of the supersaturation. Their growth rate can be assumed to be governed by the same interaction of transport and surface incorporation processes as for macroscopic crystals, and thus the rate of formation of critical-sized crystallites can be estimated.

The difference in the theories arise in the main from the different methods of estimating the free energy of an aggregate as a function of its size. The earliest developments were concerned with the nucleation of liquid droplets in vapor, in which the critical embryo appears to be of the order of one hundred particles. This is large enough to make the Gibbs approximation to the particle energy reasonable. In this approach, the energy of the aggregate is considered to be the sum of two terms: the energy of a bulk structure having the same size as the aggregate, plus a term to account for the extra free energy arising from the fact that the surface structures are different from the bulk and in general are such that a particle is more easily dissociated from the surface than from the interior of the aggregate. Gibbs equated this difference to the surface free energy of the aggregate. Although this can be done rigorously in an abstract formulation, it is by no means easy to identify this surface property with directly measurable quantities, especially at solid-liquid interfaces. By what today would be called a dimensional argument, Volmer (1934) and Becker and Döring (1935) deduced that the concentration of critical embryos should have the form

$$R_N = A \exp (\Delta G_c/kT) \qquad (1)$$

where

$$\Delta G_c = \text{free energy of the critical embryo}$$

The term ΔG_c is the sum of a surface term proportional to the embryo surface, or to the 2/3 power of the number of particles in the aggregate, plus a term proportional to the volume of the aggregate, or to the power of the number of particles. This function thus has a maximum at a value of r_c, the radius of the aggregate, or n_c, the total number of particles in the aggregate, which defines the size of the critical nucleus. At this size the free energy per particle does not change with growth. The embryo is thus stable, and those embryos to which particles arriving at a fluctuating rate will stick will become even more stable. For a saturated solution at equilibrium, the Gibbs-Thompson equation relates the size of the aggregate at equilibrium to the supersaturation of the solution. On the assumption that this relationship is equally good when there is a net rate of formation of aggregates (i.e., when the system departs from equilibrium, at least a bit), ΔG_c can be related to the supersaturation to give for the nucleation rate:

$$R_N = A \exp(-16\pi\sigma^3 v^2/(3k^3t^3(\ln S)^2)) \quad (2)$$

where v = molecular volume of the bulk aggregate
       σ = surface tension of the aggregate
       T = temperature

The pre-exponential factor, A, is the

product of two terms, the rate of growth of a critical embryo, multiplied by the number of aggregates of all sizes in the mixture. The exponential factor, as in all equilibrium theories, gives the fraction of possible aggregates which are in fact of critical size. For liquid droplets, Lothe and Pound (1962) presented a revised statistical mechanical estimate of A which was 10^{17} times larger than that estimated by Becker and Doring. Lothe and Pound's method does not involve an estimate of the value of the surface tension. Experimentally, the vapors examined appear to fall into two classes, one of which (11 compounds) follows Becker-Döring theory, and the other (6 compounds) which follows Lothe and Pound's estimates (Pound, 1972). A 15% reduction of the surface tension estimates would bring Becker and Döring's theory into agreement with the experiments on the second group of compounds, for which Lothe and Pound's formulation gives good results. The reason why the pre-exponential factor is so different between the two groups of compounds is still not understood.

In any case, statistical mechanical theory is of little help for the pre-exponential factor for crystallization from solution. We must rely on fitting experimental data to evaluate the parameters of the nucleation rate theory. Moreover, it is likely that for some crystals the critical embryo may contain relatively few particles, of the order of one or two unit cells of the lattice. In this case, a different approach has been proposed. If one considers embryo formation as a two-body addition of a particle to an aggregate, one can write a kinetic mechanism very much like that for vinyl polymerization (Becker and Döring, 1934):

$$A_{n+1} = A_n + A \qquad (3)$$

For small critical embryos, one can solve this set of transient equations numerically as a function of embryo size to determine the onset of rapid growth. This has been done by Courtney (1962) and Abraham (1969) for water vapor condensation, for

which the parameters are fairly reliably known. If one assumes that each
equation of the set (3) for $n < n_c$
reaches equilibrium, an expression
equivalent to the Becker-Döring equation can be derived (Becker and Döring,
1935). However, for very small aggregates, it is possible to use equation
(3) directly with binary reaction
kinetics at each step. In this case
one obtains a nucleation rate proportional to the n-th power of the supersaturation, where n is the number of
particles in the critical embryo. The
so-called power laws of nucleation
thus have a direct interpretation when
n is not too large.

However, there are several possible interpretations which are almost
impossible to distinguish by kinetic
experiments alone. In one model the
activation free energy for the addition of a particle is supposed to fall
rapidly with aggregate size as the
critical size is approached. This is
equivalent to the Becker-Döring use of
the Gibbs-Thompson equation to relate
surface free energy to supersaturation.
A second model would suppose that the
surface incorporation is so fast that
transport to the interface is limiting,
with diffusion proportional to the
supersaturation of the bulk solution.
The immediate cases, in which surface
diffusion and surface incorporation
have comparable rates (in some cases
even comparable to bulk diffusion
rates), could account for almost any
observed apparent kinetic order for
over-all nucleation and growth. Fortunately, the atomic processes are well
substantiated by direct observation of
the growth spirals using the recently
improved techniques of optical phase
contrast microscopy and image processing applied to in situ crystal
growth experiments (Tsukamoto, 1983).
For some time to come in technological
developments, it will be necessary to
adapt these basic concepts to the
testing and interpretation of explora-
tory experiments in the course of
developing new crystallization processes.

CRYSTALLIZER MODELLING

The study of atomic and molecular
processes on single crystal faces has
produced a fairly clear qualitative
picture of the events that must be
considered in modelling the crystallization process for single crystals.
However, many commercial applications
of crystallization depend upon control
of the crystal size and size distribution. Perhaps the most common mode of
crystallization in the chemical process
industries is production of a suspension of crystals in mother liquor. To
model this, one must consider the nucleation and perhaps the growth of a
crystal as a random process, although
subject to statistical regularities,
and influenced by process conditions
such as the magma density, suspension
residence time, and mixing of the
crystal suspension with pregnant
liquor, as well as the supersaturation
of the mother liquor.

For modelling purposes, the many
commercial types of crystallizers may
be idealized initially to two: batch,
and continuous stirred tank. In
either case, the material balance is
written for each species in the mixture, as well as for the components
of the mother liquor. Each crystallite is characterized by a single
parameter, L, called the "size", and
a balance equation is written for the
number of crystals, $n(L)dL$, in the
size range, L to L+dL, in a differential volume element of the suspension.
This balance consists of two types of
terms. The first accounts for the
transport and growth processes that
bring crystals into and out of the
size range in question without creation or destroying particles. The
second type accounts for the processes
by which particles in the size range
are created and destroyed. Creation
is by nucleatiom of critical embryos,
by agglomeration of two or more growing
crystals, or by fracture or abrasion
of a crystal in suspension. Destruction of a crystal of a given size
occurs when the crystal agglomerates to
another and loses its identity in
creating a new crystal of a larger
size, or when a crystal is fractured
or abraded to produce two smaller
crystals. The so-called population
balance can be written

$$\partial n/\partial t + \nabla \cdot (\underline{v}n) + (Gn)/L - B + D = 0 \quad (4)$$

The first two terms represent the convective transport of particles of size L,dL into the volume element. The third term represents the net accumulation of particles of size L,dL by growth on existing particles. The terms B and D represent the accumulation of particles of size L,dL by nucleation of particles of size L,dL by nucleation, agglomeration, or fracture (B), and removal from the size L,dL by agglomeration, abrasion, or fracture (D).

This result has a curious history. The steady state case for a continuous stirred tank crystallizer was first reported in the chemical engineering literature by McMullen (1935) but was not widely employed. Independently, in the early thirties R. B. Peet, working for the American Potash Company, developed the population balance for an ideal continuous stirred tank crystallizer and used it in his consulting practice for many years without publishing it (Peet, 1953). Substantially the same analysis was developed by Bransom, Dunning, and Millard (1949) in a study of the rate of crystallization of cyclonite. The work of Saeman (1956) was an early published application to commercial continuous crystallizers. The first general treatment was published by Randolph and Larson (1962). Having been introduced to the problem by R. B. Peet, Hulburt was able to refine the formulation and generalize it to the discussion of any system of identifiable "particles", be they crystals, polymer molecules, or even reactor residence times (Hulburt and Katz, 1964). Ramkrishna and Borwanker (1973, 1974) connected these ideas to a sound base in the statistical mechanics of irreversible processes, pointing out that the usual formulation of the population banace refers in fact, not to the fluctuating distribution of crystal numbers in a volume element, but to the mean value of the number density over a suitable period of time. While the fluctuations of the number density may be significant cases in which the number density is very low, in most process applications they seem to be safely ignored.

These general ideas build on those of Maxwell and Boltzman as developed in

the kinetic theory of gases, in which the "particles" are indeed actual molecules and the distributed property of interest is the molecular momentum. A "generalized "particle" is any discrete entity that is characterized by a definite set of properties, such as position in space, particle size, momentum, chain length, or residence time, for example. For each of these characterizing properties, one must know or assume an "equation of motion", i.e., the time rate of change of that property in an isolated particle. For the above list, these would be equations for the velocity, growth rate, acceleration, chain growth rate, or rate of growth of residence time, each to be expressed as a function of clock time and the other characterizing properties of the "particle".

In this framework, the growth function, G, in eq. (1) must relate the linear growth rate of the crystal to the properties which characterize the field in which a crystallite finds itself. As a minimum, these are the supersaturation of the mother liquor, the crystal's own size, and possibly the magma density. In principle, this field might also include the velocity pattern and concentration gradient near the growing surface. One must then consider the fact that different surfaces grow at different rates and that the crystal shape will also be a factor. To include these properly would require expanding the distribution function to include a shape factor. Alternatively, the shape factor might be related to the size in the absence of agglomeration or other sudden changes in habit.

None of these refinements to the population balance have yet been made in practice. Instead, perhaps guided by single crystal growth studies, an empirical expression is proposed for the growth rate. Its parameters are evaluated by fitting the resulting model to experimental data. A similar treatment is used for the nucleation and agglomeration birth function, B, and the "death" function, D. The utility of this approach depends, of course, on the success with which the parameters can be transferred from the apparatus in which they were determined to another whose performance one wishes to model. We now turn attention to these concerns.

NUCLEATION MODELS FOR STIRRED SUSPENSION CRYSTALLIZERS

Perhaps the most significant consequence of the extensive work on the theory of homogeneous nucleation is the realization that it does not account for the dominant nucleation processes in suspension crystallizers producing crystals of reasonably soluble materials. Nucleation and growth occur in these cases at supersaturations well below the predictions of homogeneous nucleation theory. It is now quite well-established that collision of crystals with each other and with the walls and agitator of the crystallizer vessel is a dominant producer of nuclei for further growth in suspension crystallizers. Long suspected by process operators, this mechanism was demonstrated by Strickland-Constable, (1966; Lal, Mason and Strickland-Constable, 1969). In this work nuclei were generated by sliding a seed crystal over a fixed plate. The effects of direct impact of crystal on crystal or crystal on other material was investigated by Clontz and McCabe (1971) by mounting the crystal in a stream of mother liquor so that it could be made to strike an anvil with controlled force. Any nuclei formed would grow to visible size down stream in the mother liquor, where they could be detected and counted. This and smaller techniques have been used by a number of subsequent investigators.

Although the importance of particle-particle interaction is well established, the nature of this interaction is still not clear. Larson and co-workers (Rusli, Larson, and Garside, 1980) have shown that attrition is probably a dominant mechanism in a stirred tank, but that the attrition rate and the size distribution of the initial fragments is a sensitive function of the rate of growth of the crystal, and hence of the supersaturation. Estrin and co-workers (Jagganathan et al.,(1980) have used an apparatus in which fluid shear, but not impact, generates secondary nuclei. As in the case of impact, they find the nucleation rate to be sensitive to the supersaturation, showing a threshhold value below which nucleation is not observed. However, it is not clear whether the shear field is producing fragments from the initial seed or simply raising the surface supersaturation to a point that fragile dendrites form.

Several investigators have suggested that crystalline clusters may form in solution near a growing face. Upon impact by another solid, these clusters may disperse and serve as nuclei for new crystals. Alternatively, they may be dispersed by shear in the fluid flowing past the crystal face, as in the Estrin or McCabe experiments. Mullin and Leci (1969) present evidence of such clusters in saturated citric acid solutions, where the high viscosity inhibits the rapid dispersal of aggregates, if they exist. Much more important work will be required to distinguish between these probabilities.

GROWTH RATE EXPRESSIONS FOR SUSPENSION CRYSTALLIZERS

The dominant factor in the growth rate is the supersaturation, but much debate has occurred over the influence of crystal size per se. While single crystal studies have shown a mild or negligible effect of the size of the face on the linear growth rate, in a suspension the concentration gradient will be affected by the slip velocity of the crystal, and thus by its size. When growth is influenced or limited by the transport of solute to the interface, one might expect an apparent change in growth rate as the size increases. With a distribution of sizes in suspension, it is a practical impossibility to model this effect in detail. The best recourse is to fit data to an empirical function that can mirror and quantify the qualitative observations.

The effect of supersaturation is probably best modelled by the predictions of the BCF theory as adapted

by Bennema, Gilmer, and their colleagues. When surface diffusion is
taken into account, this predicts a
second order dependence on supersaturation at low supersaturation, changing
to first order at higher values. However, some higher order dependence is
frequently observed in suspension crystallization. Non-integer apparent
order for growth is also commonly reported. As we have seen in the discussion of nucleation for this case, the
hydrodynamic and transport situation is
extremely complex, and it is perhaps
not surprizing that the empirical relationships are not ideally simple.

The expression

$$G = g_o(1 + g_oL)^b$$

due to Abegg et al (1968) approximates
the BCF expression, Eq. (2), over a
limited range of variables. When the
parameters G_o, g_o, and b are determined
empirically, many of the effects of
factors neglected in the model can be
implicitly accommodated. However, the
effect of scale-up on these factors
will still be unknown without further
work in which a series of crystallizers
of increasing scale are studied.

Garside and Shah (1980) have reviewed the applicability of the common
empirical expressions for nucleation
and growth in MSMPR crystallizers. It
is commonly assumed that the nucleation
and growth rates are both expressible
as simple power functions of the supersaturation, and, hence, that the nucleation rate is expressible as a power
function of the growth rate (Randolph
and Larson, 1971). Such a relationship
is very useful when the supersaturation
is very low and difficult or impossible
to measure. Comparison of data in
laboratory cooling or evaporative crystallizers for some 20 years shows a
striking consistency, despite a few
exceptions, to the empirical validity
of the relationship

$$B^o = K_R M_T{}^j G^i$$

in which the exponent i shows much less
variation between investigations on a
given system than do the nucleation
and growth rates separately. It thus
appears to be a useful empirical parameter for design. However, the accesible range of variation in the
residence time, magma density, and
supersaturation in laboratory and
pilot experiments is not large. Consequently, the precision with which
parameters such as   can be determined
is low. In addition, the systems
studied so far have usually been
chosen for ease of observation and
operational convenience. This may account for the rather narrow range of
growth rates (1 - 10 x 10^{-8} m/s) reported in these 20 systems. Experiments become either tedious (if growth
is too slow) or difficult (if too fast)
outside this range.

One must agree with Garside and
Shah that, while the population
balance formulation is a sound basis
for empirical models of crystallizers,
such models must be used with extreme
care in design and scale-up. They are
perhaps most useful in their present
state as a rough simulation to exhibit trends in the effects of modifying design or control variables and
as a guide to empirical tests in
diagnosing performance problems.

GROWTH RATE DISPERSION

An additional factor in the
modelling of suspension crystallizers
is the observation that not all crystals of the same size grow at the
same rate in the same mother liquor.
This phenomenon is most obvious in
batch crystallizers in which a uniform
change of seeds produce a non-uniform
distribution of product crystals. Recently, Garside and Ristic (1983) have
measured the growth rates of small
crystals of ammonium dihydrogen phosphate (ADP) at low supersaturations
for which the growth should be limited
by surface integration processes.
Growth rates in both x and z directions
were measured by direct microscopic
observation of the crystals during
growth. For the 56 crystals observed,
the growth rates show a distinct dispersion, representable by either a
gamma distribution or a log-normal
distribution with a coefficient of
variation of about 40% in each direction.

Such a dispersion, on the BCF theory, could reflect a dispersion in the dislocation density on the growing face, since the seeds were produced by an initial subcooling. Alternatively, dispersion has been ascribed to two-dimensional nucleation of growth centers on the crystal face. At the low supersaturation values in Garside's work this seems less likely, however. The recent advances in optical phase contrast microscopy (Tsukamoto, 1983), which will permit in situ observation of the rate of advance of growth spirals, should be very helpful in determining the range of conditions under which the various growth mechanisms are significant.

In studies of secondary nucleation, it has been observed that below about $5\,\mu$ in size accumulate more rapidly in the magma than do larger sizes and at rates that increase with the supersaturation. This has been ascribed alternatively to the possibilities (1) that small crystallites grow more slowly than larger ones, or (2) that if secondary nuclei are formed at or below the size of critical nucleus, a high supersaturation will produce smaller critical nuclei and hence a larger number of small crystals will survive during the growth phase (Strickland-Constable, 1972). The observations of Garside et al. (1979), however, suggest strongly that 1-5 micron particles can be formed by attrition in secondary nucleation, and that they grow at a slower linear rate than do larger particles. One might speculate that in the small fragment the stresses which induced fracture are substantially relieved, whereas they may persist in the larger fragment, leaving it with a higher density of dislocations, and hence a higher growth rate, than in the small fragment.

The population balance formalism was modified by Janse and de Jong (1976) to include the growth rate, G, as a distributed variable. The balance equation was then integrated over the growth rate distribution to give an equation for n(L) as a function of the average growth rate, $\bar{G}$, on the assumption that the distribution in G obeys a gamma distribution. This introduces two additional parameters in the empirical expression for G which must be determined from experiment. In favorable cases, the the precision of size distribution determinations may be great enough to permit reliable estimation of these additional parameters, but one is inclined to think that such cases may be relatively rare.

SOURCES OF DESIGN DATA

From an engineer's point of view, a model of equipment performance is of little use unless values for the model's parameters are available, either from the literature or from reasonably straightforward experimentation. Prior to the late sixties, such data were non-existent. The design engineer had to rely on the largely qualitative tests a vendor could perform in a pilot unit. The translation of these test results into design was part of the vendor's proprietary stock in trade. Fortunately, a continuous suspension crystallizer is one of the few cases in which an increase in equipment size is more apt to eliminate operating problems than to introduce them. Nevertheless, this reliance on experience and vendor's reputation was and is fraught with hazards. Today, for some types of commercial equipment the population balance formulation has provided a model for the commercial as well as for laboratory equipment which permits parameters derived from laboratory tests to be used, with judgment, in the modelling and design of commercial equipment. The mixed suspension, mixed product removal (MSMPR) laboratory or pilot crystallizer can give information on the trends in the functional relationships between nucleation and growth rates and the controllable parameters of the process. While this information is useful in understanding the operation of, say, a commercial unit of the Oslo type, with internal recycle, it should not be expected that one can make a quantitative prediction of the crystal size distribution in the larger scale unit. Neither should one expect to model cooling crystallizers well using data from evaporative units, for example. However, within the same type of hydrodynamic constraints, some transfer may be hoped for. Measurements of growth rates in a fluidized suspension of uniform seeds, for example, agree reasonably well with those measured in a laboratory MSMPR vacuum crystallizer (Rosen and Hulburt, 1971).

The ideal MSMPR crystallizer should yield a crystal size distribution of the form

$$\ln n(L) = \ln (B/G) - L/(GT_1) \qquad (6)$$

where L = Crystal size
       B = Nucleation rate
       G = Growth rate
       T_1 = Residence time

Thus a plot of ln n(L) vs. L should be a straight line from the slope and intercept of which G and B/G can be determined. For crystal sizes larger than about 100 μ, this relationship is very often obeyed. However, for smaller sizes, the plot is often observed to curve. Many explanations have been advanced for these observations, all of which yield models that have the observed curvature. These include hypotheses about size dependent crystal fracture or other secondary nucleation, size dependent growth rates, and growth rate dispersion, as well as other more esoteric possibilities. Clearly, the crystal size distribution alone is insufficient to enable one to distinguish between these very different possibilities (Garside and Jancic, 1976).

Each hypothetical explanation or model has different scale-up characteristics. Neither is there any reason to suppose that all crystals nucleate and grow by the same mechanism. This indicates the wisdom of extreme caution in using laboratory data in design unless the comparability of the test equipment and the full scale equipment has been demonstrated for the system in question.

Batch crystallizers have been used on the laboratory scale to study the rate of nucleation. Since in this case, all nuclei that survive and grow to countable size are retained in the vessel, the over-all nucleation rate is easily obtainable from measurement of the number density of the magma as crystallization proceeds. One must, of course, somehow compensate or correct for the change in supersaturation as solids are deposited if the deposition is appreciable. Many batch systems show an induction period after the initial supersaturation is established before visible crystals appear. It is often assumed that during this period embryos are growing to critical size and that at the end of the induction period, due to growth on the nuclei, the supersaturation falls rapidly to a point at which further nucleation does not occur. On these hypotheses one may set the nucleation rate equal to the number of crystals appearing divided by the length of the induction period. Nyvlt (1970, 1976) has reported such measurements on a large number of systems and has deduced power-law expressions for the dependence of nucleation rate on supersaturation. The population balance equation, Eq. (1), can be applied to this case (Hulburt, 1975) to take into account continuous nucleation and growth of crystals during the induction period. Applied to Nyvlt's data, this results in somewhat smaller exponents for the supersaturation dependence of nucleation rate, but a definitive assessment of the validity of this approach has yet to be made.

RETROSPECT AND PROSPECT

Perhaps the most significant feature of the history of industrial crystallization in the past twenty years is the increasing interplay of science and engineering in crystallization process technology. The field is enormously complex. As in many such cases, the only recourse of the engineer has been to forge ahead to the best of his knowledge and ability. Thus crystallizers in the past can hardly be said to have been designed, but have nevertheless been built and operated successfully, albeit with pain and suffering. As Rankine put it in his inaugural address as the first academic professor of engineering (1855), the scientist typically asks the question: "What am I to think?" whereas the engineer asks "What am I to do?" Solid state physics and physical chemistry provide the engineer with a sound basis for understanding the

essential features of the crystallization process. But the present state
of the science leaves much specific
detail unknown or unexplained and is
an inadequate basis for action. The
detailed knowledge which the engineer
needs in order to design, build and
operate a crystallizer must still come
from well-designed experiments on the
system of current interest, organized
and interpreted in the light of the
best scientific knowledge available.

The recognized utility of the
population balance as a framework for
interpretation of experiments as well
as the design and simulation of crystallizers makes it all the more necessary to extend our knowledge of the
nucleation and growth processes.
Techniques for direct observation of
the growing surface must be perfected
and exploited. Means of observing the
submicron particle regime are needed to
deepen our understanding of the elementary processes of nucleation and
growth. The recent advances in the
scientific knowledge of fracture
mechanics should be brought to bear on
the crystal contact processes important
for contact nucleation.

A wider variety of systems must be
explored. Most of the engineering emphasis in the recent past has been
directed to the crystallization of
fairly soluble materials. The analytical chemists have been much concerned
with precipitation of rather insoluble
substances and the "ripening" of the
precipitates. However, the engineering
of this type of system is only now beginning to receive attention (Winzer
and Emmons, 1976; Skrivanek et al.,
1976). Many organic compounds, with
their intricate molecular structure,
present different circumstances for
nucleation and growth than does the
typical soluble salt, such as an alum
or simple sulfate. Crystallization in
very heavy magmas presents challenging
modelling problems both for some industrial processes and for some environmental problems, such as the muck in
a stream bed or lake bottom in which
solid sediments may form.

Current models for nucleation and
growth, with some recent exceptions
(Bennema 1983), treat the crystal as an
equant body, a sphere or cube, with a
single characteristic length for which
one seeks the distribution. Shape is
absorbed in the ubiquitous "shape
factor", for which no predictive algorithm or theory exists for any but
the most ideal cases. Some of the
methods of quantitative taxonomy, al-
lometry, and pattern recognition should
be examined for applicability to the
characterization of shape and rough-
ness.

While in principle the crystal
property distribution in the population balance is a function of all the
pertinent characterizing properties,
in fact very little has been done for
the case of more than one distributed
property. There are basic mathematical studies on ways to handle multi-
variant systems which should be
adapted to these cases. Agglomeration
and fracture are universal events in
crystallizers, but these processes
have been incorporated only formally
in the system models.

The adaptation of the available
models to crystallizer control will
not proceed rapidly until more effective and rapid means are invented for
sensing the critical process control
variables. On-line determination of
the size distribution is presently
practical only under limited circumstances. Measurement of supersaturation, especially in the low range
needed for rapidly growing crystals,
poses a continuing challenge. Yet
effective control depends upon measuring the sensitive response variables
with adequate frequency and precision.

The development of design models,
already a lively activity which space
does not permit us to review in this
paper, will become increasingly significant in engineering practice as
our understanding grows and our experimental skills become ever sharper.
Thus it is, as Rankine observed (1855),
that "the engineer or the mechanic,
who plans and works with understanding
of the natural laws that regulate the
results of his operations, rises to
the dignity of a Sage."

LITERATURE CITED

Abegg, C. F., Stevens, J. D., Larson, M. A., 1968, A.I.Ch.E. Jour., **14**,118

Abraham, F.F., 1969, J. Chem. Phys., 51, 1969

Bransom, S. H., 1949, Dunning, W. J., and Millard, B., Disc. Farad. Soc., **5**, 83 and 96

Becker, R., and Doring, W., 1935, Ann. Phys. (Leipzig), 24, 719

Becker, R., and Doring, W., 1949, Disc. Farad. Soc., No. 5, Crystal Growth, p. 55

Bennema, P., Kern, R., and Simon, B., 1967a, Phys. Stat. Sol., 19, 211

Bennema, P., 1967b, J. Crystal Growth, 1, 278, 287

Bennema, P., 1968, J. Crystal Growth, 3/4, 331

Bennema, P., 1969, J. Crystal Growth, 5, 29

Bennema, P., and Haneveld, H. P. K., 1969a, J. Crystal Growth, 1, 225, 232

Bennema, P., and Van der Eerden, J. P., 1983, J. Crystal Growth, 61, 45

Bravais, A., 1866, A Etudes Crystallographiques, Gauthier Villars, Paris

Burton, W. K., Cabrera, N., and Frank, F. C., 1951, Phil. Trans., A243, 299

Clontz, N. A., and McCabe, W. L., Chem. Eng. Prog. Symp. Ser., 1971, 67, No. 110, 6

Courtney, W. G., 1962, J. Chem. Phys., 36, 2009

Curie, P., 1885, Bull. soc. franc. Miner

Frenkel, J., 1945, J. Phys., Moscow, 9. 392

Frenkel. J., 1946, Kinetic Theory of Liquids, Clarendon Press, Oxford

Garside, J., and Jancic, S. J., 1976, A.I.Ch.E. Jour., 22, 887

Garside, J., Rusli, I. T., and Larson, M. A., 1979, A.I.Ch.E. Jour., 25, 57

Garside, J., and Ristic, R. I., 1983, J. Crystal Growth, 61, 215-220

Garside, J., and Shah, M. B., 1980, Ind. Eng. Chem. Process Des. Dev., 19, 509-514

Gibbs, J. W., 1878, Collected Works, Vol. 1, p. 325, Reprint, Dover Publications, Inc., New York (1961)

Gilmer, G. H., Ghez, R., and Cabrera, N., 1971, J. Crystal Growth, 8, 79

Gilmer, G. H., and Bennema, P., J. Crystal Growth, 1972, 13/14, 148

Hulburt, H. M., 1975, Chemie-Ing-Techn., 47, 373-375

Hulburt, H. M., and Katz, S., 1964, Chem. Eng. Sci., 19, 555

Jackson, K. A., 1979, in Crystal Growth: a tutorial approach, ed. by Barddsley, W., Hurle, D. T. J., and Mullin, J. B., North-Holland Publishing Co., Amsterdam, p. 139ff

Jagannathan, R., Sung, C. Y., Youngquist, G. R., and Estrin, J., 1980, A.I.Ch.E. Symp. Ser., 76, No. 193, 90 and references cited there

Janse, A. H., and de Jong, E. J., 1976, in Industrial Crystallization, Ed. by J. W. Mullin, Plenum Press, New York

Kossel, W, 1927, Nachr. Ges. Wiss. Gottingen, 135

Kossel, W., 1928, in Falkenhagen, Quantentheorie und Chemie, p. 46 Leipzig

Larson, M. A., and Berglund, K. A., 1982, A.I.Ch.E.. Symp. Ser., 78, No. 215, 9

Lal, D. P., Mason, R. E. A., and Strickland-Constable, F. R., 1969, J. Crystal Growth, 5, 1

Lothe, J., and Pound, G. M., 1962, J. Phys. Chem., 36, 2080

Mason, R. E. A., and Strickland-Constable, R. F., 1966, Trans. Faraday Soc., 62, 455

McCabe, W. L., 1929, Ind. Eng. Chem., 21, 30 and 112

McMullen, R. B., 1935, Trans. A.I.Ch.E., 31, 409

McMullen, J. W., and Leci. C. L., 1969, Phil. Mag. 19, No. 161, p. 1075; J. Crystal Growth, 1969, 5, 75-76

Nernst, W., 1904, Z. Physik. Chem., 47, 52-55

Noyes, A. A. and Whitney, W. R., 1897, Z. Physik. Chem., 23, 689-692

Nyvlt, J., 1970, Industrial Crystallization in Solutions, Butterworths, London

Nyvlt, J., 1976, "Design of Batch Crystallizers", in Industrial Crystallization, J. W. Mullen, Ed., Plenum Press, New York, p. 335

Ohara, M. and Reid, R. C., 1973, Modeling Crystal Growth Rates From Solution, Prentice-Hall, New York

Peet, R. B., 1953, Personal communication to the Author

Pound, G. M., 1972, Natl. Bur. Stds. J. Phys. Chem Ref. Data, 1, 119, quoted in F. F. Abraham, Homogeneous Nucleation Theory, Academic Press, New York, 1974, p. 105-107

Ramkrishna, D., and Borwanker, J. D., 1973, Chem. Eng. Sci., 28, 1423

Ramkrishna, D., and Borwanker, J. D. 1974, Chem. Eng. Sci., 29, 1711

Randolph, A. D., and Larson, M. A., 1962, A.I.Ch.E. Jour., 8, 639

Randolph, A. D., and Larson, M. A., 1971, Theory of Particulate

Processes, Academic Press, New York, p. 46

Rankine, W. J. M., 1855, On the Harmony of theory and practise in mechanics in Applied Mechanics, 11th edition, Charles Griffin & Co., London, 1885

Rosen, H. R., and Hulburt, H. M., 1971, Chem. Eng. Prog. Symp. Ser., 67, No. 110, p. 18, 27

Rusli, I., Larson, M. A., and Garside, J., 1980, A.I.Ch.E., Symp. Ser., 76, No. 193, 52 and references cited there

Saeman, W. C., 1956, A.I.Ch.E. Jour., 2, 107

Skrivanek, J., Zacek, S., Hostomsky, J., and Vacek, V., 1976, in Industrial Crystallization, e. by J. W. Mullen, Plenum Press, New York, p. 173

Stranski, I. N., 1928, Z. Phys. Chem., 136, 259

Stranski, 1949, Disc. Farad. Soc., No. 5, Crystal Growth, p. 13

Strickland-Constable, R. F., 1972, A.I.Ch.E. Symp. Ser. 68, No. 121, 1

Tsukamoto, K., 1983, J. Crystal Growth, 61, 199-209

Verma, A. R., 1953, Crystal Growth and Dislocations, Academic Press, New York

Volmer, M., and Flood, H., 1934, Z. Physik. Chem., A170, 273

Volmer, M., 1939, Kinetik der phasenbildung, Dresden

Winzer, A., and Emons, H.-H., 1976, in Industrial Crystallization, ed. by J. W. Mullen, Plenum Press, New York, p. 163

Wulff, G., 1901, Z. Krystallogr., A100, 272

AZEOTROPIC AND EXTRACTIVE DISTILLATION

Donald F. Othmer ■ Polytechnic Institute of New York, Brooklyn, NY 11201

Mixtures of two liquids vary as a continuum from those which are completely miscible over a wide temperature range, through all degrees of partial solubility as the mixtures become less ideal, soluble or compatible; to those which are substantially insoluble. The composition of the vapors also vary as a continuum and may be calculated simply only for those mixtures at either end of the scale:—for ideal, completely soluble pairs, by Raoult's Law; or for quite insoluble pairs, by additive vapor pressures, steam distillation. In some ranges of this continuum the composition of the vapors will approach that of the liquid then, at some point the vapor will have the same composition as the liquid, at a lower boiling point than either of the pure liquids.

This phenomenon, *azeotropy*, is used in distillation processes to separate a first liquid from the other by *increasing its relative volatility*, thus lowering its boiling point to that of the azeotrope. Many azeotropes, on condensation, separate into two liquid layers:—the first is separated, the other then is recycled to separate more of the first liquid.

In extractive distillation, a high boiling solvent descends the distilling column, preferentially extracts the higher boiling liquid, *lowers* its relative volatility thus increasing its effective boiling point and removes it as bottoms. Wide industrial usage has applied both methods since 1900.

THEORY

Vapor Liquid Relations

All distillation practice is based on the relation of the composition of vapor in equilibrium with and arising from a boiling mixture of two or more liquids. (A binary solution is meant here unless otherwise indicated). Vapor-liquid relations for binary mixtures of liquids may be calculated readily and precisely only for truly ideal pairs or for truly non-ideal pairs:

1. Ideal solutions are of liquids which are very closely related chemically and physically, such as adjacent members of an homologous series; and have perfect solubility, miscibility, and compatibility, in every sense of these words, (e.g., n-hexane and n-heptane, wherein a molecule of one may substantially replace a mole of the other in the liquid's molecular configuration). Vapor compositions of such ideal solutions may be calculated by Raoult's law.

2. Absolutely non-ideal pairs are those of liquids which are practically completely non-compatible, non-miscible, non-soluble, etc., such as hexane and water. With such mixtures the total vapor pressure is calculated precisely by the familiar addition of the vapor pressures of the two pure liquids at the given temperature. If water is one, this gives what is called a steam distillation. Molar vapor composition is then directly proportional to the respective vapor pressures of the individual liquids.

A Continuum -- A Phenomenon in Continuous Change

In between these two extremes of binary mixtures there is a continuum of types of mixtures, and indeed of vapor-liquid relations: since at one specific total pressure of a distillation, the vapor-liquid relation of the same binary mixture may be considerably closer to the ideal end of the continuum than at another total pressure. Different pairs of liquids are known which fill every possible gradation of this continuum between almost perfect ideality or solubility and almost perfect non-solubility or non-ideality.

This relation of the vapor to liquid compositions of different binary mixtures is indeed related to the solubilities of the two liquids. Thus while methanol and water are usually regarded as completely soluble at ambient temperatures, this may be only in the

Presented at: - AIChE - 75th Anniversary Meeting, Nov. 1, 1983 Symposium on: - History of Fractional Distillation

90

normal liquid temperature range; and
ice separates out of or is not completely soluble, in, say, a solution
of 20% methanol - 80% water at -15°C.
Conversely to this separation out of
solution at low temperatures, many
binaries which are only partially miscible at ambient temperatures become
completely miscible at a high consolute
point - although this would certainly
not be an ideal solubility or compatibility.

Thus, Figure 1 shows the familiar
vapor-liquid relations or x,y plot for
an ideal binary solution having almost
absolute mutual solubility or compatibility. The system of hexane and hep-
tane gives x,y relation almost exactly
the same as that calculated from the
mole fractions of the liquid compositions multiplied by the respective
vapor pressures of the pure components
at the boiling temperature and divided
by the total pressure of the system;
i.e., the system follows Raoult's law.

The vapor composition relations
of the non-ideal and insoluble mixture
of the system of hexane and water are
also shown in Figure 1; this time as a
horizontal line of constant vapor composition and a two-phase condensate
from the two insoluble liquids boiling
together throughout the entire range
of mutual insolubility of the liquids
of this boiling point. This is constant
as long as there are two liquid phases
in the still. This line of constant
vapor composition crosses the x = y
diagonal line to give what is called
an azeotropic point, in this case, a
heterogeneous azeotrope where the
boiling liquid has the same ratio of
the two components as the condensate.
The ends of this horizontal line are
determined by the respective solubilities of hexane in water and of water
in hexane at the boiling temperature
of the two-phase mixture. Each of
these values is practically zero; and
is much too small to be indicated
exactly in Figure 1 because of the
thickness of the lines.

In between the two pairs of liquids
whose vapor-liquid relations are represented as the two extremities of the
continuum in Figure 1, there are all
grades of dissimilarity, immiscibility,
non-solubility or incompatibility of
systems of two liquid components. Calculations and predictions of the compo-

sitions of vapors rising from such
intermediary systems are not possible
by such simple laws or techniques as
Raoult's law for hexane-heptane, or
the law of additive partial pressures
of immiscible liquids, for hexane-water.

Selection of different binaries
is readily made to show that there is
a greater or lesser divergence from
ideality or compatibility or solubility
which is a difference of degree rather
than a difference of kind. The terms
"completely miscible" or "forming two
phases" are roughly indicative of
solubilities or compatibilities; but
the lesser than complete mutual compatibility of liquids may show up more
dramatically in the vapor composition
relations than in the solubility relations with which they are closely
associated.

Thus, Figure 2 plots the vapor
composition at atmospheric pressure
(in this case always plotted as mole
percent water) of the systems made up
of water with the normal alcohols containing from 1 to 5 carbon atoms.

There is a major gradation in the
shape of the x,y curve since the alcohol
molecule becomes more and more like a
water insoluble hydrocarbon as the
number of carbon atoms increases to
overbalance the effect of the hydroxyl
group. Water with methanol approaches
the ideal at atmospheric pressure.
However it does deviate at higher
pressures. Water with ethyl alcohol
is not quite so near to ideality and
forms an azeotrope (the best-known one)
at 10.7 mole % water. Water with propyl
alcohol also gives a homogeneous azeotrope at 56.9 mole % water. Water with
n-butyl alcohol gives two phases due to
limited solubility. This is indicated
by a horizontal line, since the composition of each of the two liquid phases
must be constant during a boiling-off
operation, although the ratio of the
amounts of the two phases, and hence
the over-all composition of the liquid
is changing. The azeotrope is where
the horizontal line crosses the diagonal
at 76.8 mole % water. Water and normal
amyl alcohol have less mutual miscibility; and thus there is a longer
horizontal line showing a constant
vapor composition over a wide range of
liquid composition and an azeotrope at
85.8 mole % water.

All of these are minimum boiling

azeotropes, and their boiling relations may be more or less indirectly influenced by the additive pressure phenomena of steam distillation, or the approach thereto of liquids which are more or less compatible. Even though methyl, ethyl and propyl alcohols are completely miscible with water, they form aqueous systems which are less and less completely compatible as the molecule becomes more like that of a hydrocarbon. For example, even ethyl alcohol and water are not completly soluble or compatible at low temperatures, where water becomes insoluble and crystallizes out of the solution.

Possibly a more dramatic illustration of the continuum of types of vapor-liquid relations is indicated by the single system: acetone-water at different pressures. Here is a binary system of liquids which is completely miscible, not necessarily completely compatible; and the differences or incompatibilities may show up much greater at some pressures and temperature than at others; more often at higher pressures and temperatures.

Figure 3 shows the x,y curve for acetone and water at one atmosphere pressure to be slightly concave upwards in the range of higher acetone compositions. This shows a first noticeable deviation from ideality as demonstrated by the x,y curve. At higher pressures, this tendency causes the curve to dip progressively lower, until at a pressure of about 40 lb./sp. in. abs., it crosses the x = y line, indicating an azeotropic mixture. Figure 3 shows the azeotrope to decrease in content of acetone with increasing pressures (1). This tendency to form an azeotrope under conditions when the system is of completely miscible liquids is typical of other systems at other temperatures and pressures. The system of water and acetic acid (the only volatile organic acid-water system which does not form an azeotrope) is quite far from ideality and has a concavity upward; also acetic is partially associated to a dimer in the vapor phase, At higher pressures, the x,y curve drops, as does that of acetone-water, but the critical region is reached without the curve crossing the diagonal to indicate an azeotrope. (2).

Another case, the well studied ethyl alcohol-water azeotrope at atmospheric pressure has long been known to disappear at a pressure of approximately 0.1 atm., and at lower pressures the concavity upward appears.

Azeotrope Mixture Formation

Keyes (3), later Ewell, Harrison and Berg (4), and more recently Berg (5) and various others have discussed the many variations of liquid systems and the formation of azeotropic mixtures as a function of the ideality of the mixture based on the tendency to form hydrogen bonds. Francis (6) discussed the degrees of mutual compatibility or solubility of liquids, and classified, with graphs, liquids as having: (a) complete miscibility, (b) high solubility, (c) low solubility, and (d) very low solubility. As shown above the completely miscible class should be broken down still further on the basis of what might be called compatibility to indicate even lesser divergence from ideality which shows up when vapor phase relations are also considered.

Usually, the boiling points of materials forming homogeneous azeotropes are not far apart; but heterogeneous azeotropes may be formed by liquids of quite different boiling ranges (e.g. water and various immiscible liquids) where the vapor pressures are strictly additive. These are also constant boiling mixtures since the vapors are constant over a wide range of liquid compositions, the whole range of immiscibility.

The ends of the horizontal line of constant vapor composition which is the usual mark of a heterogeneous azeotrope, are determined by the solubilities of the respective liquids in each other at the temperature of the constant boiling mixture. In Figure 1 for the benzene-water system, the horizontal line extends practically the entire range of liquid compositions because of the nearly complete insolubilities of these two liquids, while for butyl alcohol and water as in Figure 2 the much greater mutual miscibilities stops the horizontal line at the points of saturation, as shown where the curves break off the horizontal line at both its left and its right extremity

In some very few cases, such a horizontal line may not cross the

diagonal x = y line. There may be a
large range of liquid composition
wherein the vapor composition is constant and this represents, from the
standpoint of the vapor, a constant
boiling mixture. However, this range
may not include that point where the
liquid composition itself equals this
vapor composition. Thus, there is no
azeotrope on the horizontal line. One
interesting example, the system methyl-
ethyl ketone and water, has been plotted
at different pressures in Figure 4 (1).
At each of the x,y curves for a lower
constant pressure, there is a wide
horizontal band of boiling at one temperature and one fixed vapor composition - but varying liquid composition.

The solubility limits are indicated
by the extremities of these horizontal
sections of the x,y curve, which form
an interesting solubility diagram,
dotted in Figure 4. In every case,
there is also a homogeneous azeotrope
to the right of the solubility limit
as defined by the horizontal line, and
in a range where there is a complete
miscibility in boiling and condensation.
At pressures where the boiling temperatures are above that of critical solubility (the minimum point of the dotted
curve) no horizontal line can appear;
and the x,y curve of the system becomes
that of the usual binary of two completely miscible liquids with a minimum
azeotrope.

Maximum Boiling Azeotropes

Azeotropes may also have a maximum
boiling point. They are always homogeneous, and are comparatively few in
number but are of interest theoretically.
They may give problems in separation
in industry. Typical examples are
those of water with highly ionized acids
(nitric, hydrochloric, formic) wherein
an association, hydration, or related
loose bonding probably forms entirely
different molecular species from the
fundamental components, water and acid,
with correspondingly different vapor
pressures. Such systems are also known
among systems without an acid; but
there may be somewhat similar molecular
functions involved. Formic acid and
water are difficult to separate, but a
modification of the processes herein-
after discussed for dehydration of
acetic acid solutions (7) is quite good.

THERMODYNAMIC CORRELATION

The thermodynamic background of
azeotropy as a function of total pressure, and also as a function of
temperature, may be demonstrated simply
by considering that: 1. the vapor
composition in terms of moles of one
component equals the ratio of the
partial pressure of that component to
the total pressure; and 2. at the azeotrope, where vapor composition equals
liquid composition, the activity
coefficient of either component is the
ratio of the total pressure to the
vapor pressure of that component at
the azeotropic temperature. (8)

$$y_1^A = p_1^A / P \qquad (1)$$

and

$$\gamma_1^A = P / p_1^O \qquad (2)$$

Here y_1^A is the vapor composition
(molar), γ_1^A is the activity coefficient
and p_1^A is the partial pressure of the
one component in the azeotrope, and
p_1^O is the vapor pressure of the same
component in the pure state; all at
the azeotropic temperature at the total
pressure P.
These equations may be expressed in
logarithmic form, respectively:

$$\log y_1^A = \log p_1^A - \log P \qquad (3)$$

and

$$\log \gamma_1^A = \log P - \log p_1^O \qquad (4)$$

Both equations 3 and 4 may be differen-
tiated with respect to $\log p_1^O$,

$$(d \log y_1^A)/(d \log p_1^O)$$
$$= (d \log p_1^A)/(d \log p_1^O)$$
$$- (d \log P)/(d \log p_1^O) \qquad (5)$$

and

$$(d \log \gamma_1^A)/(d \log p_1^O)$$
$$= (d \log P)/(d \log p_1^O)$$
$$- (d \log p_1^O)/(d \log p_1^O)$$
$$= \left[(d \log P)/(d \log p_1^O) \right] - 1$$

However, the term

$$(d \log P)/(d \log p_1^{\,o})$$

is the slope of the line on logarithmic paper of the plot of total pressure, P, against the vapor pressure $p_1^{\,o}$ of a reference substance, in this case one of the pure components. This line is known to be straight (9, 10); hence the slope is constant; and the term

$$(d \log \gamma_1^{\,A})/(d \log p_1^{\,o})$$

(being equal to a constant minus unity) must also be constant. This is the slope of a plot on logarithmic paper of the activity coefficient vs. the vapor pressure of a reference substance always taken at the same temperature. The slope is constant, hence <u>the line must be straight</u>.

Similarly, the composition $y_1^{\,A}$ of the azeotropic vapor can be shown to give a straight line when plotted on logarithmic paper versus the vapor pressure of a reference substance. This is shown for the 18 systems of Table 1 in Figure 5. Fundamentally, Figure 5 is a plot against temperature; the temperature scale being specially developed thermodynamically to give straight line plots (9). A simpler plot, not depending on a reference substance, was developed for activity coefficient and vapor composition plotted against the total pressure (10). On logarithmic paper, these may be shown to be straight lines by the same reasoning, as shown also in Figure 6 and 7 for the systems of Table 1.

A further combination of thermodynamics and analytical geometry may be used (11) to predict vapor-liquid equilibrium data and to develop linear equations and straight lines on logarithmic plots for azeotropic - as well as other - mixtures at conditions of either constant temperature (as preferred by the physical chemist) or of constant pressure (as preferred by the chemical engineer). The data required are of temperature-composition and either latent heats or vapor pressures of the pure components.

From the basic Gibbs-Duhem equation, there can be developed (11) for binary systems (and similar ones for ternary systems): -

$$\left(\frac{x}{y} - \frac{1-x}{1-y}\right) \cdot \left(\frac{dy}{dx}\right)_P = \frac{L_x}{RT^2} \cdot \left(\frac{dT}{dx}\right)_P$$

This equation resembles the Redlich-Kister equation - but does not have its assumptions, and has somewhat more general usage and applicability. Stepwise integration gives the vapor composition curve, i.e., Figure 8, for alcohol-water, the best known system with an azeotropic mixture. If in this equation there is substituted the values Y for y/(1-y) and X for x/(1-x), straight line plots and linear equations will represent Y vs. X for all binary mixtures, and with suitable modification, all ternary mixtures. Ideal solutions give on logarithmic paper single straight lines as shown in Figure 9 with a slope of 45°; non-ideal solutions, including all azeotropic mixtures, give three straight lines, one for each end of the x,y curve and one for the middle section.

The above derivation from the Gibbs-Duhem Equation is thus a fundamental tool for predicting equilibrium data of azeotropic as well as other mixtures. Also it is useful for correlating and checking experimental vapor-liquid equilibrium data at a constant pressure. Thus, by substituting in its left side the proper values of x and y, and in the right side the corresponding boiling point and latent heat of vaporization, substantially identical values for the many systems worked with have been obtained, which ensures thermodynamic consistency of the experimental data. This is shown by the straight line obtained at a 45° angle for every such plot made (11). Another method of checking the x,y,T data available for a system is to plot (dT/dx) calculated from the equation against dT/dx obtained experimentally. Excellent checks were obtained for the many azeotropic and other systems tested (11).

By slight further modification and steps, vapor-liquid data for binary or ternary systems - see Figure 10 - can be predicted or correlated with a thermodynamic rigor which depends only on Dalton's law of partial pressures and the Clausius-Clapeyron Equation. The mathematical interrelation of the thermodynamic functions of liquid mixtures, both ideal and those giving azeotropes such as: - temperature of boiling points, vapor pressure, latent heats, vapor-liquid equilibria, activities, relative volatilities, solubilities, etc., has been demonstrated to give straight lines geometrically and simple linear functions algebraically. When applied with the additional capabilities and large memories of large computers, or even hand calculators, simple programs are obtained which give almost immediately through their great power the correlation and/or prediction of data for one - or more - of these values because of its intimate interrelation with all of the others. Such correlations and predictions can be made with confidence under vacuum, atmospheric pressure, or moderately higher pressures.

COMMON SEPARATIONS BY AZEOTROPIC DISTILLATION IN THE CHEMICAL INDUSTRY

It is often difficult or practically impossible to separate industrial solutions of two or more liquids by ordinary distillation and rectification, because of the closeness of the boiling points; i.e., the x,y curve is very close to the diagonal x = y; also , in many cases, where an azeotrope is formed. Here the x,y curve crosses the diagonal indicating that, at the particular pressure of the distillation, the azeotropic mixture acts, from the vapor pressure standpoint, as a single compound; and no separation beyond it can be achieved.

Azeotropic distillation is used as a separating tool for such liquids; and several of the azeotropic distillation systems used in industry indicate the possibilities of this technique. The simpler ones may be considered first, rather than the first one historically, which happened to be less simple.

Water must often be separated in industry from another volatile liquid. This may use a distillation with an entrainer (also called a withdrawing agent) which is a liquid which forms with water, in this case a heterogeneous azeotrope having a lower boiling point than any other component of the mixture. Figure 11 diagrams this typical and simplest azeotropic separation. It

was first described in 1927 (12, 13) as a basic flow sheet for water removal with an entrainer, and for many other systems. It is also obvious that water may be added as an entrainer for one of some pairs of organic liquids with which it azeotropes; and the converse of the following operation would be used for the separation of such a pair.

The aqueous binary mixture (the usual and simplest example) enters by line A at the top of the column if the second liquid is saturated with water, or by line B at an intermediate point if less than saturated. Entrainer may be charged in advance to the system through line C. In the left, or dehydrating column, the water and entrainer are distilled to give the azeotropic vapors overhead as the lowest boiling constituent of the liquid system. The azeotropic vapors are condensed to give two liquid phases (a heterogeneous azeotrope) which flow to a decanter, where they are separated. The entrainer layer is passed back to the top plate of the dehydrating column as reflux to distil with, or entrain, more water; and the counter-current rectifying action of the column brings up and over the top the azeotrope in practically the theoretical proportions, since it is the most volatile component of the mixture being distilled. This reflux stream is now desirably "all of part" of the vapors which have now been separated by condensation substantially into the two immiscible liquids; practically all of one of which is removed as reflux for the dehydrating column, depending on the degrees of insolubility. On the other hand, the reflux of the usual distillation system is "part of all" the condensed vapors since there is no separation of the constituents of vapors leaving the column on condensation.

The liquid being dehydrated passes downwardly in the column, becoming dryer and dryer, finally discharging practically water-free at the base, which is heated by closed steam.

The water layer leaves the decanter saturated with the entrainer. The dissolved entrainer is removed and returned to the dehydrating column to repeat its entraining action there. Hence this water layer is stripped in a distillation which brings over the top of the water column all of the

dissolved entrainer in the <u>same azeo-</u>
<u>tropic vapor mixture</u>. This is passed
to the same condenser since it has the
same composition as the vapors from the
dehydrating column. The condensate
is the same two liquid phases which
goes to the decanter; and the process
contiues indefinitely. The water is
separated by the entrainer in the
dehydrating column, a trace of entrainer
remains, and the water contains less
and less of the entrainer as it descends
the water column. It is finally passed
out of the bottom almost pure. (Open
steam may be used for heating the base
of this column containing this pure
water; if other material is present, a
closed heating coil is used.)

Water in a small amount, dependent
on its solubility with the entrainer,
continually cycles around the top of
the system; but usually this is too
small in amount to effect the equipment
size or steam consumption. Thus, in
an exemplary system, the azeotrope
might contain 20% water, 80% entrainer,
the water might dissolve 1% of the
entrainer, and the entrainer might
dissolve 2% of water. One pound of
water is thus brought over by 4 pounds
of entrainer which, in being refluxed
to the dehydration column, would carry
back about 0.08 pounds water. About 1
pound of water layer, in its turn,
carries about 0.01 pounds entrainer
back to the water column. This might
be distilled out of the water column
in the azeotrope with 0.04 pounds of
entrainer. The water column is thus
seen to be very small, as only about
1% as much vapors leave it as leave
the dehydrating column.

In this idealized system, the
reflux ratio is <u>fixed by the ratio of</u>
<u>entrainer to water in the azeotrope.</u>
To change this, the entrainer may be
changed. In practice, if more reflux
is needed than is given by the particular entrainer, some of the water
layer may be returned by a line not
shown in Figure 11.

Drying Butanol and Other Self Entrainers

A by-product of World War I and
its needs for acetone to dissolve
nitrocellulose was butanol, which is
produced simultaneously in the fermentation of carbohydrates to give acetone.
After nitration and washing of nitro-

cellulose, butanol displaces and dissolves the last of the wash water. A
water layer is decanted from the saturated butanol and handled separately.
The butanol layer, was, for years,
distilled for water removal in a simple
pot still without rectification. Much
more butanol came over than is in the
(heterogeneous) azeotropic mixture at
93°C; and the temperature was much
higher because of non-approach to
equilibrium. The decantation was
extremely slow because the final fractions near the solubility limits,
contained more butanol than the azeotrope and very small droplets of water.
This, one of the simplest examples of
azeotropic distillation, was developed
in 1927 as the flow sheet of Figure 11,
to reduce the many thousands of gallons
of butanol then in decantation storage
because of the low settling rate, and
to reduce the excessive steam cost
from the large number of redistillation

Here, in Figure 11 butanol, saturated with water, is fed to the top of
the dehydrating column by line A. The
water azeotropically distills with the
butanol itself. No difficulty is
experienced in rectifying to obtain,
as vapors at the top, the azeotrope
at 93°C containing 55% butanol and
45% water,

The condensate from the vapors
decants rapidly to give an upper layer
of about 72 vol. % butanol and a lower
layer of about 28 vol. % water, each
layer saturated with the other component. This is about 22 wt. % water in
the upper layer, and 8 wt. % butanol
in the lower layer. The continuous
decanter refluxes the entire upper
layer of butanol back to the dehydrating
column, where there is only one liquid
phase - butanol - saturated with water
at the top and decreasing to dryness
at the bottom. The fractionation is
rapid and efficient, as shown by the
steepness of the equilibrium curve of
the water-butanol system on the left
of the horizontal line in Figure 2.
The number of plates is immediately
calculable based on usual methods, and
the reflux heat of the butanol layer
being returned. This reflux is quite
adequate to hold down butanol in excess
of the azeotropic mixture. Only a
relatively few plates give dry butanol
at the base.

The water layer leaves the decanter

saturated with butanol. It passes to the top of the water column, also a stripper. Here, again, the reflux is adequate to give water free of butanol at the base of a relatively short column, while the same azeotropic vapors go overhead. This column may be calculated from the curve on the right of the horizontal line for water - butanol in Figure 3. Open steam may be used for heating the water at the base. The vapor streams from both columns are the same composition, the azeotropic; and they are passed to the same condenser, which delivers directly to the decanter. A major saving is obtained in steam and equipment and particularly in holdup or storage of butanol in the system.

Drying Hydrocarbons - also Emulsion Breaking

During the last twenty years, many processes have been described to dry hydrocarbon of their relatively small amount of dissolved or entrained water (14) substantially by the process developed for butanol in 1927.

One current development is the azeotropic removal of water from emulsions with petroleum crudes stabilized by ultra-fine solid particles so as to be practically permanently dispersed (15). Particularly, the emulsions of crudes from tar sands are difficult to separate; and when high speed centrifugation is used, the erosion and hence destruction of the expensive machines has added a large cost to that of producting the crude. The small amount of water may be entrained at a low heat cost, by lower boiling hydrocarbons which may be present, or added if necessary.

DEHYDRATING ACETIC ACID

The use of acetic acid increased greatly for the production of cellulose acetate for film and rayon use in the late 20's, also for other processes. These gave large quantities of dilute acid, which had to be concentrated. Ordinary rectification even in a column with many plates, has a prohibitive heat cost to obtain water at the top, nearly free of acid, from acid at the bottom, nearly free of water. It is economically impossible to prevent substantial loss of acid at the top, although dry acid can be obtained at the bottom with a reasonable amount of reflux.

The same flow sheet of Figure 11 was developed (12, 16, 17), after various other processes with more steps were tried. The aqueous acetic acid, often in a vapor state, was fed more nearly to the middle of the dehydrating column, i.e., line B, Figure 11.

However, in dehydrating acetic acid, an entrainer must be added to bring over the water. The entrainer effectively decreases the boiling point, or increases the effective vapor pressure, of one of the components; in this case water. The dehydrating column is charged through line C, with an entrainer which, by forming an azeotrope with the water, reduces its boiling point.

Ethylene Dichloride as Entrainer

Using ethylene dichloride, the effective boiling point of water drops from 100° to $71.6^{\circ}C$. This was the first entrainer used commercially for this purpose (12, 16, 17). The azeotropic vapor condenses to form two layers, which are decanted. The entrainer forms in this case the lower layer since ethylene dichloride is heavier than water. It is passed back to the column; and the water layer is refluxed to the water column.

Acetic acid boils at $118^{\circ}C$ at the base, substantially higher than the effective boiling point now of water. An excessive of entrainer is used, and mixtures of dry acid and entrainer boil at intermediate temperautres in the lower part of the column; and the azeotrope, 92% ethylene dichloride and 8% water, goes off the top at $71.6^{\circ}C$. The water layer in the decanter contains about 0.8% dissolved ethylene dichloride, and the ethylene dichloride layer contains about 0.2% dissolved water.

The water column acts as before to strip out the very small amount of entrainer which is dissolved in the practically acid-free water. It is heated with open steam at the base, where water discharges containing about 0.1% acetic acid. A very slight loss of ethylene dichloride over long continued operation requires some make-up through line C.

Propyl Acetate as Entrainer

In 1932, propyl acetate was found to be better than ethylene dichloride; but since its boiling point, $101.6^{\circ}C$, is even higher than that of water, any excess over that required for the azeotrope would be rectified from acetic acid in the lower part of the column only with great difficulty. Hence, it was attempted to maintain exactly the right amount of propyl acetate in the acid column to form the azeotrope. Thus at the point in the column where the last of the water distilled out of the aqueous acid, the last of the entrainer formed the azeotrope with it (18, 19, 20). The greater water content in the azeotrope (14 wt. %, compared to 8% for ethylene dichloride), enables the same column still to produce a much greater amount of product, at a much lower unit heat cost than when using ethylene chloride.

Butyl Acetate as Entrainer

Butyl acetate was later found to give even better results; it has a higher azeotropic boiling point $90.7^{\circ}C$ with 27% water in the azeotrope (21, 22). Any excess of butyl acetate in the lower part of the acid column, and below the level of descent of the water, cannot be separated from the acid in this single column, since it has a higher boiling point $126.3^{\circ}C$, than that of acetic acid, $118^{\circ}C$. To guarantee against this, insufficient butyl acetate to form an azeotrope with all of the water is present in the lower part of the column. Thus the liquid descending the column contains water below the point at which the last of the butyl acetate has been vaporized off.

The rectification of water containing some acetic acid at the bottom of a column is not difficult; there is no pinch in the tray calculations between the lower end of the equilibrium curve and the operating line based on the amount of reflux back from the decanter, even with butyl acetate, which involves much less reflux heat than with propyl acetate, which, in turn, involves much less reflux heat than with ethylene dichloride. Hence, a bottoms product can be produced above

99.5% acid continuously, with a much greater capacity of the column and a much lower steam consumption. The water column operates very much as before, with water discharging containing only 0.01 to 0.1% of acetic acid. Design methods have been described for these comparative systems (22).

The loss of entrainer may be quite small - for example, in long term industrial operations, per ton of acetic acid dehydrated, only 1 to 2 pounds of butanol may be added, and esterified to butyl acetate in the system.

Comparison of Entrainers

In this operation of the flow sheet of Figure 11, the reflux in the dehydrating column is fixed by the ratio of the particular entrainer to water in its azeotrope, or more exactly by the ratio of the two liquid layers leaving the decanter. Entrainer passing into the top of the acid column does this here. This is well illustrated in Figure 12.

The normal function of reflux wash is the condensing from the rising stream of vapors of the less volatile constituent, which is acetic acid, and dephlegmating it down the column. If a change in the reflux ratios is required there must be a change of azeotropic ratios; and this means a change of entrainers.

However, since in much of the upper part of the dehydrating column, there are two liquid phases, the entrainer is extracting acetic acid from the aqueous phase into the entrainer liquid phase.

Thus the function of knocking down or dephlegmation of the acetic acid, i.e., reducing its volatility, depends upon the partition coefficients or the extracting efficiency compared to water of the acid-free entrainer refluxing to the top of the column. Hence, while ethylene dichloride produces much greater reflux wash than does propyl acetate, and very much more than does butyl acetate, its very low partition coefficient when in contact with water for acetic acid means that it is very much poorer in dephlegmating the acetic acid than would be calculated from the respective reflux wash availabilities. The upper part of the column is extractively distilling, and ethylene dichloride is a very poor extracting solvent for the acetic acid.

The choice of an entrainer depends on: (a) boiling temperature, this controls the ratio to water in the azeotropic mixture, (b) the partition coefficient with water of acetic acid, as well as (c) the concentration of the dilute acid to be handled. Even higher boiling entrainers than butyl acetate have been successfully used particularly with dilute acid.

Also an entrainer, such as propyl acetate may be used first as a solvent for a preliminary extraction of the acetic acid; then by passing the extract layer to a distilling column, the water is azeotropically removed to give dry acid, and some of the decanted entrainer is paased to the extractor, while the balance goes, as usual, to be reflux wash to the top of the dehydrating column.

DEHYDRATING ALCOHOL

Dehydrating butanol represents the simplest example of azeotropic distillation as a self-entrainer; and acetic acid is more involved by requiring an added entrainer which forms an azeotropic mixture with only one of the liquids being separated.

Ethanol (alcohol) presents a more difficult problem in that most entrainers form binary azeotropes with both water and alcohol - and usually a ternary azeotrope of all three compounds. However, the need for 100% alcohol (absolute) and the great usage of distillation techniques in the alcohol industry, made alcohol the first liquid to be dried by azeotropic distillation even with the greater problem of all three liquids forming azeotropes.

While water boils at 100°C, Young (23) noted in about 1900 that the ethyl alcohol-water azeotrope boiled 21.7° lower, at 78.3°C for ethyl alcohol, compared with 100°C for water, and contained 95.6 wt. % ethyl alcohol and 4.4% water. This was the highest concentration which could be reached by ordinary distillation; and it is always the desired feed strength for any system to make absolute alcohol. Up to Young's time, chemical methods of dehydration were used. On the other hand, he found the heterogeneous binary azeotrope of benzene-water boiled at 69.4°C, which was 9°C lower than the alcohol-water azeotrope and contained 91.1% of benzene and 8.9% of water.

Young reasoned that the bezene-water azeotrope, because of its lower boiling point, could be rectified overhead from absolute alcohol at the bottom of a column. He did find that the benzene would bring over the water in an azeotrope at the top of the column; and give absolute alcohol at the base. Unfortunately, however, this was as a ternary azeotrope at the top - the first noted - between benzene, ethyl alcohol, and water. It boils at 64.6°C, and contains 74.1% benzene, 18.5% alcohol, and 7.4% water.

The ternary azeotrope came in a first fraction when distilling the three liquids. This was separated into two layers. Then came a dry, alcohol-benzene fraction; then an absolute alcohol fraction. Water was added to the first fraction to separate a water layer with much of the alcohol. This was withdrawn for concentration of the alcohol by usual distillation. The benzene layer of the first fraction was chemically dehydrated with quick lime; and it, with the second fraction was returned to the next batch operation.

Young's basic batch distillation process requiring chemical treatment was improved by many workers, over many years into several economic and continuous processes using benzene (24). Usually these required additional columns over the basic flow sheet of Figure 11, because of the necessity of separating the ternary mixture of entrainer, alcohol, and water from the aqueous layer of the decanter. The best benzene process by Katzen (25), has many plates in the azeotropic column, with no benzene present in the lower ten plates, after the technique using insufficient entrainer in dehydrating acetic acid (21, 22).

During this same time, chemical engineers searched in vain for an entrainer of suitable physical properties which does not form an azeotrope with alcohol, thus to keep alcohol out of the overhead vapors and condensate. It was not until 1940 that one was found (26).

Ether forms an azeotrope with

water, but <u>not with alcohol</u>, possibly because it is so closely related to and compatible with alcohol. Ether was not considered, however, because of its low boiling point; thus the amount of water in its azeotrope is very small. At atmospheric pressure, this azeotrope boils at 34.15°C, and the composition is 98.75% ether and only 1.25% water. Less water dissolves in ether at lower temperatures, so decantation at the lowest practical temperature is desirable: at 15°C ether dissolves about 1.15% water. Thus, only an extremely small amount of water (about one part in a thousand) would be decanted in the flow sheet of Figure 11.

By increasingly the pressure in the dehydrating column (26), the amount of water in the azeotrope goes up considerably (as noted also in Figure 3 for acetone-water azeotropes and in Figure 4 for methyl ethyl ketone-water azeotropes). At about 8 atmospheres, water represents over 4 wt. % of the vapors leaving the columns, ether is less than 96%. If the decanter of Figure 11 operates at 15°C, the water layer is about 3 wt. % of the azeotrope.

Only a small amount of water is to be separated. The latent heat of ether is low, so it is practical to make absolute alcohol by distilling as much as 1.5 to 2 times as much entrainer as feed of 95% alcohol.

The very small water column and decanter also may be operated at the higher pressure of the dehydrating column. Thus, the process operates with all of Figure 11 at the elevated pressure; but the decanter is at as low a temperature as the cooling water allows so as to minimize the solubility of water in the ether layer which is returned to the azeotropic column.

The temperature of this azeotropic distillation now has become higher than that of other distillation operations in producing alcohol from the fermentation liquors; and the heat given up in the condenser may be used again by the Vapor Reuse Method (27, 28) for operating other of the distillations, by what is a super-double-effect evaporation.

The Vapor Reuse Method

The vapors arising from the top of a beer still fed with a usual mash containing about 8% alcohol will give a condensate of about 40% alcohol. The rectification of this condensate to high proof spirits will require much less heat than that given upon its condensation.

The Vapor Reuse System utilizes this principle by operating the beer still at an elevated pressure so that its vapors can condense in the reboiler of a combined exhausting and rectifying column. This is boiling substantially pure water to give steam which condenses in the reboiler of the alcohol column, operating at a lower pressure to give sufficient temperature drop for heat to transfer to boil the almost pure water at the base of the alcohol column. The larger amount of heat than is necessary for accomplishing the rectification of the condensate from the beer still may be used, if pure alcohol is to be made, in operating a "heads" column to remove undesired aldehydes and other low boilers.

Because of the arrangement with different pressures, this is somewhat akin to a double effect evaporator, but it has the additional advantage of utilizing the principle mentioned above, depending on the vapor composition relations.

Figure 13 illustrates the usual Vapor Reuse Process with vapors from the beer still, which operates under pressure, being passed for condensation first to the reboiler of the alcohol column and a heads column, both usually operating at atmospheric pressure. Even accounting for the heat requirements of these two columns, there is an excess of heat available as low-pressure steam from the reboiler of the alcohol column. Condensate from each column is separated by its own trap; and because of the pressure which is substantially that of the beer still, each stream flows to a mid-plate of the heads column which takes off volatile, undesirable heads. The alcohol and water leave the bottom of this heads column and are pumped to a mid-point of the alcohol column.

Alcohol of 95% concentration goes overhead, and the water is discharged from the base of the alcohol column. (Separation of fusel oil is not indicated, since this may be done as standard practice).

Several advantages have been found, however, if the beer still is operated at <u>atmospheric pressure</u> instead of at a higher pressure. Then the alcohol column and the heads column, using the heat of these vapors, are operated at a <u>partial vacuum</u> to give the necessary temperature difference (29).

In designing a major plant, it is usually desirable to be able to make <u>not only alcohol for motor fuels</u>, but the <u>highest possible</u> purity product for use in the chemical, pharmaceutical, food and other industries when gasohol is superseded by cheaper fuels; and the alcohol is to be sold and used as has been conventionally, particularly also so that it may compete with synthetic alcohol.

A very small amount of quite volatile impurities may be difficult to remove from the concentrated alcohol produced by the alcohol column. In practice, concentrated alcohol has often been diluted again to a low concentration to increase the volatilities of the impurities to be brought over in a steam or azeotropic distillation before taking off the alcohol, which is purified significantly by this treatment. The improved system using Vapor Reheat, which is much less expensive of heat, now has a different arrangement of equipment and several "wash" trays on top of the beer still supplied with a small amount of hot distilled water from a later part of the process to separate these undesirable impurities.

The stillage water is discharged from the base of the beer column; and the water, separated from the 35 - 40% vapors, is discharged from the base of the rectifying column. The heat in both streams is recovered by conventional heat interchanging with the fermented mash entering the system. As usual, there is the conventional recovery of other values in the stillage from the beer column.

ABSOLUTE ALCOHOL PRODUCTION FROM BEER

Alcohol in motor fuels is usually dissolved in from 80 to 95% gasoline. Anhydrous alcohol must be used since small amounts of water causes a separation of two liquid layers in the gasoline tank. The lower one, largely water, is quite incombustible.

Conventional distillation gives a high concentration of about 95 - 96% alcohol, which is a good fuel but too expensive in this country. Thus an azeotropic, dehydrating distillation is required to remove the last 4 or 5% of water, using an entrainer which steam distils out this small amount of water to remove it from the resulting anhydrous or absolute alcohol.

Dehydrating Column

Azeotropic entrainers - most often benzene - withdraw the water, but form azeotropic mixtures with the alcohol as well as with the water. <u>Ether</u> does not, and thus allows a simpler dehydration system. The use of ether makes it possible to eliminate, down to parts per million, the entrainer from the product alcohol, and ether is much less objectionable in most uses of the absolute alcohol produced than the higher quantities of benzene, always remaining. The amount of water in the azeotropic mixture increases significantly with the higher pressure which is using about 13 atmos. and $130^{\circ}C$.

With the substantial difference between the boiling point of the azeotrope of ether and water at the top of the column and that of pure ethanol at the bottom, the separation is easy, and the alcohol discharges from the base of the dehydrating column above 99.85% in concentration and with less than 10 ppm of ether.

A diagram of the molar and hence the approximate heat quantities in the Dehydrating Column is shown in Figure 14 to illustrate its operation. Here widths of bands represent the number of moles flowing (alcohol, water, ether) in both liquid and vapor. Most of the water brought overhead in the azeotropic mixture is separated in the Decanter, saturated with ether. This ether is removed by passing the water to a Water Stripping Column. This distils out the ether in the same azeotropic mixture. Its vapors, being the same, are combined with those from the Dehydrating Column.

Combined Flow Sheet

The block diagram of Figure 15 is only a flow sheet, drawn to <u>no scale</u>.

Some conventional items are not included, such as: Liquid-Liquid Heat Exchangers, Valves and Controls, Vent Connections and Vent Condensers, Steam and other Condensate Traps, etc. (The plant diagrammed produces 150 million ltrs/yr. of absolute alcohol of either a pharmaceutical grade or a fuel grade.)

In general, the heat flow is from the boiler steam to heat both the Dehydrating Column and the Water Stripping Column which operate under pressure. The azeotropic mixture of vapors from both are combined to pass to heat the reboiler of the Beer Still, operating at atmospheric pressure. Vapors from the Beer Still, operating on the Vapor Reuse principle, heat the reboilers of the Alcohol column and then the Alcohol-Heads Stripping Column, both of which in this design operate at a slight vacuum and are interconnected as shown.

The combined vapors from the Alcohol Column and the Heads Stripping Column are rectified in the Upper plates of the Alcohol Column and are passed to a Condenser, then finally to a vent condenser which is not shown. Residual gases are removed by a steam ejector, creating a partial vacuum. The hot water discharged in a small amount from the base of the Water Stripping Column is passed to the top one of a few plates on top of the Beer Still to give a slight wash to the vapors there, which operate as Vapor Reuse in the Alcohol Column and Heads Column.

One group of impurities, water-insoluble oils, the so-called fusel oil fraction, steam distils with the water present in the Alcohol Column at a temperature too high to allow it to go overhead with the alcohol and too low to allow it to be drawn off with the water from the bottom. An emulsion collects in the lower part of this column. This is drawn off to a decanter, or to be washed in a liquid-liquid extractor. The washed oils are separated for use; and the water, containing a small amount of alcohol, is passed back to a lower plate to be discharged also from the bottom of the column. This fusel oil separation is another example of azeotropic distillation.

Economics

Space prevents full discussion of details, exact arrangement of accessries and sizes of the different units of equipment. Slight differences of piping and flows of materials and heat have secured improved performance. Moreover, newly designed plants starting with fermentation mashes as prepared by some of the newer techniques will use less than 15 pounds of steam to make one gallon (1.8 kilograms per liter) of anhydrous, chemically pure alcohol. As in all processes, the steam consumption is lowered if, instead of a chemical pure grade of 100% alcohol, there is produced a fuel grade anhydrous alcohol or a fuel grade 95% alcohol. Using high fermentation alcohol mashes, heat consumption then will be a low of 12 pounds of steam per gallon or about 1.4 kilograms per liter. Water consumption will be small indeed because of the multiple use of heat, in the reboilers of other distilling columns, rather than in the discard of so much heat to cooling water. Equipment costs as quoted for new plants are also substantially lower than that for conventional plants of the same capacities.

SEPARATING FORMIC AND ACETIC ACID

The two lowest fatty acids are each unique in its relation with water. Formic acid boils at almost the same temperature and has a maximum azeotrope with water; acetic acid is the only volatile fatty acid which has no azeotrope with water at any pressure (2). Higher volatile members of the series are progressively less compatible and less soluble with water, thus forming azeotropes, following the pattern of the alcohols diagrammed in Figure 2.

Boiling solutions of acetic acid and formic acid give vapors very nearly the same as those of the liquid. The x,y curve is very close to the diagonal; and the separation of these two acids, which often come as mixtures in industrial processing is difficult. Usually there is also some water present. No entrainer for formic acid has been found which does not also form a heterogeneous azeotrope with water; but with some there is no ternary azeotrope of the entrainer, water, and formic acid.

Chloroform has been found to be a satisfactory entrainer for both formic acid and the small amount of water which may be present in solutions (7, 30). The azeotropes are: chloroform-water at 56°C with 97.2 wt. % chloroform; chloroform-formic acid at 59°C with 85 wt. % chloroform. There is no azeotrope of chloroform and acetic acid, and none for any three or all four of the liquids.

The aqueous mixture of the two acids (usually, principally acetic acid with minor amounts of formic acid and water) is fed by line B to the dehydrating column of Figure 11 which is charged with chloroform. A vaporous mixture of the two azeotropes, i.e., chloroform-water and chloroform-formic acid, comes to the condenser together because of their close boiling points. Practically no acetic acid, the fourth component, is in these vapors; since it forms no azeotrope with any of the other three liquids. The condensate is decanted; and the chloroform layer is returned to the dehydrating column, carrying only a small amount of water and formic acid. The aqueous layer passes to the water column carrying most of the formic acid and some chloroform. The chloroform is distilled out of the water column in the same two binary azeotropes (with water and with formic acid); and the process at the top operates the same as in other examples, with two azeotropes, however, instead of one. The water column discharges all of the formic acid with all of the water from its base, which, in this case, must be heated by closed steam rather than with open steam as in those cases where water alone is discharged.

If no water is present, the homogeneous chloroform-formic acid azeotrope is formed. Then a small amount of water, about 5 to 10% of the formic acid being separated, is added through line C of Figure 8, to cause formation of two layers in the decanter. This runs off continuously with the formic acid at the base of the water column.

Such an addition of water (sometimes another liquid) may be used to break the single phase condensate into two liquid layers with the basic flow sheet of Figure 11. One or two accessory columns will work up the liquid streams which result. A more efficient use of the water stream minimizes the amount required for the formation of two layers. Instead of the simple addition through line C, the decanter may be replaced with a countercurrent, liquid-liquid extractor, in which the single phase condensate is extracted with water to discharge an extract layer (with most of the formic acid) to the water column, and a raffinate layer (with most of the chloroform) back to the dehydrating column.

Aqueous formic acid may be dehydrated as is acetic acid, but the selection of entrainers is more difficult because of the greater similarity, affinity, and compatibility of formic acid and water, the very close boiling points, and the very poor partition coefficients for formic acid as compared to acetic acid from aqueous solutions. With a satisfactory entrainer, e.g., propyl formate (13), the process operates with the flow sheet of Figure 11, as used for dehydrating acetic acid.

SEPARATING HYDROCARBONS

The major quantities of liquids to be separated, by far, are the hydrocarbons - for fuels and lubricants, and increasingly since 1930, as chemicals or intermediates in making other chemicals. Fuels and lubricants are seldom pure compounds but are cuts by boiling ranges. Often these cuts or fractions contain one or more series of hydrocarbons of the same boiling range, some with undesirable properties. The separation is then by series, as olefins or aromatics from paraffins, rather than by individual compounds as alcohol from water. And by adding an entrainer for one whole series of hydrocarbons within the boiling range of the fraction, it may sometimes be possible to distil the compounds of the one series as their azetropes away from the other series.

Usually no entrainer liquid has sufficiently great difference in compatibilities with members of the two series to form a heterogeneous azeotrope with one series of hydrocarbons, and no azeotrope with the other. However, an alcohol may form homogeneous azeotropes with the numbers of one series in the narrow boiling range without forming any azeotropes with the members of the other. The azeotropes with the members of one series distils them away from those of the other, which are separated as a bottoms product; the azeotropes are condensed; water is added to the condensate to separate it into two layers as has been described previously for other systems. The alcohol is separated or extracted out in the water layer, and is distilled therefrom for reuse; and the water-washed hydrocarbon layer, comprising members of one of the series may be substantially free of the hydrocarbons of the other series which came out of the bottom and vice versa.

Berg (5), Keyes (3) and others (31, 32, 33) have discussed the selection of entrainers for such separations.

AZEOTROPIC OR PARTIAL PRESSURE DRYING OF SOLIDS

For many years, water in solids has been determined in the laboratory by adding an entrainer such as benzene or toluene, azeotropically distilling out the water, separating and returning the benzene. The same system has been used for drying of foods (34) for preservation.

Anhydrous caustic soda in bead form, and with some advantages in processing costs has been produced from the 50 or 70% solution coming from the evaporators, using kerosene as the entrainer (35) for the water at a temperature 275°C lower than that required in conventional fusion pots, and at a every substantial saving in energy costs. Because of its low energy cost, this process was used in Japan during World War II.

Oxychlorides often form in drying solutions of various metallic chlorides; and this also may be prevented by an azeotropic removal of the water.

If a concentrated solution of two salts is agitated to give droplets in a suitable boiling withdrawing agent (often a narrow cut or fraction of kerosene) the droplets individually lose their water and a permanently uniform mixture of the two solid components is obtained in each particle.

Similarly, if it is desired to coat crystals with another water soluble material, seed crystals of one solid are agitated in a concentrated solution of the other material; and on immiscible entrainer distils the water out azeotropically and leaves each crystal uniformly coated.

A further expansion of these processes is the simultaneous intimate mixing and coating of particles to secure an end product of great uniformity. For example, (36) particles of carbon may be coated with sulfur dissolved in an entrainer-solvent and potassium nitrate dissolved in water by agitating and distilling off the azeotropic mixture to give black gun powder. (Possibly an improvement over the classic and ancient Chinese development.)

SPEEDING CHEMICAL PREACTIONS

Esters and Soaps

Water is a common by-product of chemical reactions. In esterification of volatile acids and alcohols, the continuous removal of the water as it is formed by an azeotropic distillation encourages the reaction to go to completion by preventing the reverse reaction. This technique has been used for generations (36, 37, 38) to take advantage of the vapor-liquid relations for the volatile liquids (alcohol, acid, ester, and water) and thereby to remove water by azeotropic distillation.

Special distillation columns are used in industry with deep trays designed to retain a large hold-up. This, in continuous operations, slows the passage of the liquids flowing downwardly plate to plate and hence allows a longer time for chemical reaction between the components. Also special sections may be provided in the column wherein sulfuric acid as a catalyst may be added to speed the reaction and then be neutralized and washed out in a caustic section at a lower plate in the column. Most plant designs have been empirical. However, some studies - as those at the Polytechnic Institute of New York (37, 39, 40) - have correlated, plate by plate in the distilling column, actual operating results against values calculated from data of the reaction kinetics and the vapor-liquid equilibria. Usually the azeotroic entrainer is one of the liquid components, in other cases, an

additional liquid is added as
entrainer.

Saponification or fat splitting
is the opposite of esterification.
Here the alcohol, glycerol, is formed;
and it has an azeotrope with a suitable
low cut kerosene at about 200°C under
atmospheric pressure - lower under
vacuum. In a continuous process (36)
the insoluble glycerol and kerosene
are both completely removed to give a
high quality soap having some advantages. Greases and heavy metal soaps
may be made similarly.

Phenol by Sulfonation and Caustic Fusion

Benzene may be sulfonated in
kerosene, which raises the boiling
point to speed the reaction and
entrains the water formed azeotropically. A purer sulfonic and sodium
salt gives phenol on fusion with
caustic soda; and again, the presence
of kerosene as an entrainer to remove
the water formed gives high yields
and a more readily purified product
(41). Phenol by this dual azeotropic
process, has been made industrially
in Japan; and other related compounds:
resorcinol, p-cresol, B-naphthol, etc.,
have also been made to show the
general utility of the system.

Nitrobenzene, Nitrotoluene

Benzene (42) and toluene (43)
may be nitrated by removing the water
formed continuously, utilizing as the
entrainer the hydrocarbon itself,
rather than removing water batchwise
by the classic dilution of mixed sulfuric acid and nitric acid. The concentration of the nitric acid may thus
be maintained at the optimum value.
No sulfuric acid is used, and there
is thus no attendant problem or expense
of spent acid recovery; all of the
nitric acid fed to the reaction is
utilized in one operation. The
nitrated hydrocarbon, and an excess
of hydrocarbon, are passed from the
base heater of the dehydrating column,
which also acts as a reaction vessel,
to a lower column still. The nitrated
hydrocarbon has the higher boiling
point and is separated as a bottoms
product of this lower column, while
the hydrocarbon vapor goes to the

bottom of the dehydrating column.
This process was used in Germany during
World War II.

Optimum temperatures and concentrations may be maintained easily in
this system, with the simplicity of
control of the usual continuous distillation unit. In this case, the
variables are: (1) the rate of feed
of nitric acid and of hydrocarbons to
the reactor, (2) the rate of supply
of heat to each of the two heating
units, also (3) the rate of withdrawal
of water at the top of the dehydrating
column and of mono-nitro compound at
the bottom of the lower column. The
product is formed in a single passage
and with almost 100% yield of both the
nitric acid and the hydrocarbon.

Acetamide

Other reactions, wherein one product is water or other more or less
volatile liquid which is capable of
forming an azeotrope, may be expedited
in continuous processes through azeotropic distillation. Often a flow
sheet very similar to that for the
nitration of benzene and toluene is
used. One is the continuous production
of acetamide (44).

Continuous streams of gaseous
ammonia and acetic acid are fed to the
dehydrating column of Figure 11. Ammonium acetate is formed with one mole
of water which is continuously dehydrated in the azeotropic dehydrating
column. A second mole of water is
formed in the chemical dehydration of
the ammonium acetate to give acetamide.
The water formed in both reactions
goes overhead aided by a suitable
added entrainer. Acetamide with excess
acetic acid flows out the base of the
column to the top of a second distilling column, which separates pure acetamide at its base. The acetic acid
passes back as vapor to the base of
the dehydration column, of Figure 11,
which merely has an additional column
operating below the dehydrating column
to give pure acetamide.

Further Examples

Other azeotropic distillations
have been used for many years for
removal of the water formed in still
other chemical reactions to accomplish

the completion of the reaction.

The steam distillation of aromatic oils and other very slightly soluble, flavorsome liquids enhances the aroma of many foodstuffs from hot coffee and pastries, bacon and gravies. Also the perfumes of damp flowers will be more fragrant in the moist atmosphere of a dewy morning or of a green house than that of dry flowers at the same temperature and a much lower humidity outside. There, the vapor pressure of the essential oils is not raised effectively by that of water.

EXTRACTIVE DISTILLATION

Extractive distillation is usually considered along with azeotropic distillation for two reasons. (a) Often they are being conducted simultaneously, or at different levels of the same distilling column: - azeotropic as a part of an extractive distillation and extractive as a part of an azeotropic distillation. For example, the reflux of an entrainer which is a solvent for acetic acid, as propyl acetate or butyl acetate, to the top of azeotropic column washes or extracts the acetic acid from the vapors rising in the column and carries them down and away from the acid-free vapors going overhead. (b) However, in a sense, extractive distillation is the converse of azeotropic distillation in that the added liquid in the extractive distillation - there called the solvent - reduces the effective relative volatility of one of the components, acetic acid in its separation from water, while in an azeotropic distillation, the added liquid, there called the entrainer, increases the relative volatility of one of the components, there the water. Thus in fortunate cases, the third liquid may separate the pair by increasing the volatility of the more volatile one and decreasing that of the less volatile one.

As noted above an entrainer may be added to accomplish an azeotropic distillation (e.g., of water from acetic acid). It forms a constant boiling mixture with the water to give water a significantly lower effective boiling point and a significantly higher relative volatility; and to allow the azeotrope to distil away from the acetic acid whose effective

boiling point is much less affected. If a high boiling solvent is added, it may not affect the volatility of the water so much, but its solution with the acetic acid has a lower volatility than that of the acid alone,

A diagram of an idealized extractive distillation process is indicated in Figure 16. This, like the idealized diagram of the azeotropic process of Figure 11, almost always will have variations. These depend on the properties of the liquids involved, particularly the vapor pressures, compositions, and mutual solubilities - also on impurities, either liquids or solids which are invariably present.

An ideal solvent is assumed for A which is one of the components of the binary mixture; and A ideally has very much less solubility and a higher boiling point than W, the other component. The extractive distillation column operates the same as any distillation column with W passing as vapors overhead and to a condenser. Part of the condensate W is drawn off as one product, and a part (usually much smaller) is refluxed. The high boiling solvent is fed into the top of the column, and acts as the usual reflux. It dissolves and condenses out vapors of A to be carried to the bottom, ideally with none of W. The mixture of solvent and A passes from the base of the main column to a midpoint of a solvent stripping column which, by its conventional action, separates A overhead as the other product.

In Figure 16 a heat exchanger is indicated to preheat the incoming feed of A + W while cooling the solvent from its high boiling point at the base of the solvent stripping column. Other accessories also are usually necessary.

In considering an actual extractive distillation, such as the separation of water, W, from acetic acid, A, a relatively high boiling solvent for acetic acid is used which is immiscible with water. This solvent is added to the top plate of the column which is distilling water, W over the top from the higher boiling acetic acid, A. The solvent acts as a liquid reflux in the column because of its high boiling point. As it descends it extracts the acetic acid, A, from the rising vapor stream. The relative volatility of the acid in the liquid on the plates is <u>decreased</u> and its effective boiling point is increased.

Since there is no perfect separation by any solvent and indeed it dissolves some water, there is a mixture of high-boiling solvent, acid, A, and usually some water, W, finally discharged from the base of the column at a boiling point between that of acetic acid, A, and that of the solvent. This mixture is fed to a second column where it is separated to give 70 - 85% acetic acid (A + W) overhead, while dry, substantially acid-free solvent is removed at the bottom. The 70 - 85% acetic acid is the feed for a third column which concentrates it to 99.5+ % by usual rectification and discharges it from the base, while the water with some acid is removed overhead and then recycled.

This process was used by Suida in the 1930s for the dehydration of approximately 30% acetic acid obtained from the production of cellulose acetate. However, he had developed it (45) in 1927 for the separation of a more complicated liquid mixture: - the condensate of vapors from the destructive distillation of wood, called pyroligneous acid, and containing 6 - 7% acetic acid. Besides the many creosote and tar oils present in pyroligneous acid, a part of which Suida fractionated separately to use as the high boiling solvent for his extractive distillation, pyroligneous acid contains several percent methanol (wood alcohol).

The crude pyroligneous was vaporized to be the feed to the extractive distillation column, the water and methanol vapors rose through a wash of solvent which extracted the higher boiling tar oils along with the acetic acid as it descended in the column. The acid-solvent mixture was drawn off the bottom and separated, usually in a vacuum distillation. The acetic acid, now concentrated to 70 - 80%, went overhead in the second column, and the bottoms product - the tar oils - was recycled as solvent to the top of the extractive distillation.

Another column separated the small amounts of methyl acetate, acetone, as well as the methanol and other low-boiling, water-soluble materials distilled overhead from the extractive distillation column along with the

water vapors, also small amounts of solvent and other higher-boiling water-insoluble materials coming over in an azeotropic or steam distillation. This crude methanol column was usually equipped with side draw-offs and decanters for separating the wood oils, in the same way as fusel oil is separated in the alcohol distillation of Figure 15.

In hydrating acetic acid, compared to azeotropic distillation, extractive distillation requires a more complicated and expensive plant and more steam. Dehydration of acetic acid was also the problem which started the development around 1880, of another chemical engineering operation, continuous <u>liquid-liquid extraction</u>, which has become also an important process in separating liquid mixtures.

A concentrated calcium chloride brine was used in one of the earliest extractive distillations (46) as the solvent reflux from the top plate of a column wherein the azeotrope of methanol and acetone was boiling. Calcium chloride solutions have a strong affinity for methanol as compared with that for acetone, and alcoholates are formed - similar to hydrates - with water-soluble alcohols. A practically quantitative separation of acetone from methanol may be obtained, with acetone going overhead, while the methanol passes out in the bottoms, dissolved in the calcium chloride solution.

In a batch operation, the pot was not filled, and space was left in the pot for accumulation of the brine used as solvent for the methanol. When all of the acetone was removed overhead, the boiling temperature was increased and methanol was distilled out of the pot with some water. Methanol was rectified from this mixture in the usual manner to give a pure product. The brine remaining in the pot was pumped to a storage tank for reuse with the next batch.

A continuous operation is obtained by the use of the process of Figure 16. The brine containing the methanol as it leaves the bottom of the extractive distillation column is passed to the short stripping column which separates pure methanol overhead. The brine leaving the bottom is passed through a heat exchanger for cooling and from there as recycle to the top of the extractive distillation column.

Friedland showed later (47) that water, in a somewhat larger amount than of brine, may used as the extracting solvent to absorb and hold down the methanol while allowing the acetone to go overhead. The water is fed to a plate of the extractive distillation column above that of the feed; the plates between the two liquid feeds accomplish the water extraction of methanol, and those above the water feed separate acetone from water. Acetone is stripped from the aqueous methanol in that part of the column below the feed; and aqueous methanol discharges at the base to be separated in a second column.

A third liquid thus accomplished in these and many other cases, the separation by extractive distillation through its increase of the effective boiling point of one of the two liquids, that is the decrease of the vapor pressure and relative volatility of one liquid of the original binary mixture, preferably the higher boiling one. Often the boiling point of a liquid is raised by the addition of a salt, and hydrophilic salts have long been used to combine with water in the drying of liquids. Furter has also added a dry salt feed to the top of the distilling column to remove the water as a brine from the absolute alcohol which goes overhead (48).

Furter and co-workers have identified in numerous other articles many liquid systems which have been separated by extractive distillations using a strong salt solution as the extractive agent, as in the case discussed above of brine separating methanol from acetone; and he has listed over 500 references (e.g., 48 - 49). The binary mixture is separated because of the elevation of the boiling point, or lowering of the vapor pressure, of one of the liquids (usually water) by the salt solution, thus allowing the other liquid to pass off as a vapor   In some cases this may be called, or at least likened, to, the salting out effect used so frequently in work with aqueous solutions of organic materials. Thus the use of salt solutions has long been known to raise the effective boiling point of water in alcohol solutions to such

an extent that the water passes as a salt solution out of the base of the column while absolute alcohol leaves overhead.

Solutions of both hydrophilic salts and liquids absorb vapor from water which is being cooled by its evaporation, in absorption refrigeration cycles. This absorption is possible because the effective vapor pressure of water out of the solution is lowered. similarly, solutions of the same salts and liquids may be used in extractive distillation. For example, ethylene glycol has been long suggested as the solvent in an extractive distillation to remove water in a bottoms solution, while allowing absolute alcohol to go off the top as vapors. As in the absorption refrigeration cycle, the aqueous hydrophilic, high boiling liquid is then evaporated to remove the water, and the hydrophilic liquid is recycled to absorb or extract more water. However, the amounts of heat used are shown by Black and Ditsler to be higher than by a separation using azeotropic distillation (50).

Mixtures of liquids to be separated in recent decades have been mainly those derived from petroleum products; and many extractive distillation processes have been develped for separations in the petroleum and petrochemical industries. During this same period, mathematical modeling of the physical processes has allowed the theoretical analysis of the separation of azeotropic and other close-boiling mixtures of liquids with the assistance of computers.

Tassios has used gas-liquid chromatography as a means of selecting solvents and has given a bibliography (50) of these instrumental and computer modeling means for simulating and possibly predicting extractive distillation processes.

LITERATURE CITED

1. Othmer, D.F., M.M. Chudgar, and S.L. Levy, Ind. Eng. Chem., 44, p. 1872 (1952).
2. Othmer, D.F., Silvis, S.J., and Spiel, A., Ind. Eng. Chem., 44, p. 1864 (1952).
3. Keyes, D.B., Ind. Eng. Chem., 33, p. 1019 (1941).
4. Ewell, R.H., J.M. Harrison, and L. Berg, Ind. Eng. Chem., 36, p. 871 (1944).
5. Berg, L., Chem. Eng. Prog. 65, p. 52 No. 9, (Sept. 1969).
6. Francis, A.W., Ind. Eng. Chem., 36, p. 1096 (1944).
7. Othmer, D.F., and J.J. Conti, U.S Patent 3,024,170 (March 1962).
8. Othmer, D.F., and Ten Eyck, E.H., Ind. Eng. Chem., 41, p. 2897 (1949).
9. Othmer, D.F., Ind. Eng. Chem., 32, p. 841 (1940).
10. Othmer, D.F., and Gilmont, R., Ind. Eng. Chem., 36, p.858 (1944).
11. Othmer, D.F. et al., Ind. Eng. Chem., 45, p. 1815 (1953).
12. Clark, H.T., and D.F. Othmer, U.S. Patent 1,804,745 (May 12, 1931).
13. Clark, H.T., and Othmer, D.F., U.S. Patent 1,826,302 (Oct. 6, 1931).
14. Brown, B.T., Clay, H.A., and Miles, J.M., Chem. Eng. Progress 66, p. 54, No. 8, (Aug. 1970).
15. Balassa, L., and Othmer, D.F., Advances in Chemistry Series, Am. Chem. Soc., Washington, D.C. 1972.
16. Othmer, D.F., Trans. Am. Inst. Chem. Engrs. 30, p.229 (1933).
17. Othmer, D.F., U.S. Patent 1,917,391 (July 11, 1933).
18. Othmer, D.F., U.S. Patent 2,028,800 (January 28, 1936).
19. Othmer, D.F., U.S. Patent 2,028,801 (January 28, 1936).
20. Othmer, D.F., U.S. Patent 2,050,235 (August 4, 1936).
21. Othmer, D.F., U.S. Patent 2,050,234 (August 4, 1936).
22. Othmer, D.F., Chem Met. Eng., 48, p. 91 (June 1941).
23. Young, S., Trans. Chem. Soc., 81, p. 707 (1902).
24. Keyes, D.B., Ind. Eng. Chem., 21, p. 998 (1929) & US Pat. 1,676,735 (1928).
25. Katzen, R. (Personal Communication) November 1976.
26. Othmer, D.F., and Wentworth, T.O., Ind. Eng. Chem., 32, p. 1588 (1940).
27. Othmer, D.F., Ind. Eng. Chem., 28, p. 1435 (1936).
28. Wentworth, T.O., D.F. Othmer, and G.M. Pohler, Trans. Am. Inst. Chem. Engrs., 39, p. 565 (1943).
29. Wentworth, T.O. and Othmer, D.F. Alcohol Distillation at Low Heat Cost, (A. I. Ch. E. National Meeting, Detroit, Mich. Aug. 1981)
30. Conti, J.J., Othmer, D.F., and Gilmont, R., J. Chem. Eng. Data, 5 p. 301 (1960).
31. Brame, J.S.S., and T.G. Hunter, J. Inst. Pet. Tech., 65, 802 (1927).
32. Bruun, J.H., and M.M. Hicks-Bruun, Bur. Standards J. Research, 5, p. 933 (1930).
33. Rossini, F.D., B.J. Mair, and A.R. Glasgow, Jr. Proc. Am. Petrol. Inst. III 2/5, p. 43 (1940); also Bur. Standards J. Research, 26, p. 39, 565 (November 1941).
34. Bohrer, B.B., U.S. Patent 3,298,109 (January 17, 1967).
35. Othmer, D.F., and Jacobs, J.J., Ind. Eng. Chem 32, 154, (1940).
36. Othmer, D.F., Ind. Eng. Chem 33, 1106 (1941).
37. Berman, S., Isbenjian, A. Sedoff, and D.F. Othmer, Ind. Eng. Chem. 40, p. 2139 (1948).
38. Kirk, R.E., and Othmer, D.F., ed., "Encyclopedia of Chemical Technology," Interscience Encyclopedia, Inc., New York, 1st Edition., vol 5, p. 776 (1950).
39. Leyes, C.E., and D.F. Othmer, Trans. Am. Inst. Chem. Engrs., 41, p. 157 (1945).
40. Leyes, C.E., and Othmer, D.F., Ind. Eng. Chem., 37, p. 968 (1945).
41. Othmer, D.F., and Leyes, C.E., Ind. Eng. Chem., 33, 158 (1941).
42. Othmer, D.F., and J.J. Jacobs, Jr., J.F. Levy, Ind. Eng. Chem., 34, p. 286 (1942).
43. Othmer, D.F., H.L. Kleinhans, Jr., Ind. Eng. Chem., 36, p. 447,(1944).
44. Othmer, D.F., U.S. Pat. Appl. (1932).
45. U.S. Patents 1,624,812 (1927); 1,697,738 (1929); 1,703,020 (1929) H. Suida
46. U.S. Pat. Appl. (1927), D.F. Othmer.
47. D. Friedland, unpublished thesis, Polytechnic Institute of New York, 1945.
48. W.F. Furter and R.A. Cook, Int. J. Heat Mass Transfer, 10(1), 23 (1967).
49. W.F. Furter, Can. J. Chem. Eng., 55(3), 229 (1977).
50. Tassios, D.P., Ed. Advances in Chemistry Series-115, A.C.S. Washington, D.C., 1972.

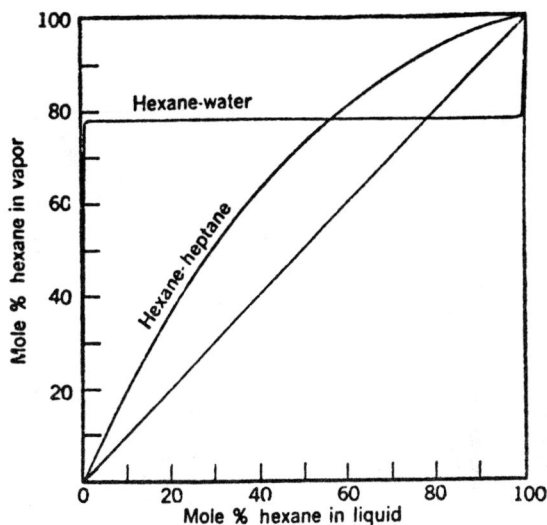

Figure 1. Vapor-liquid equilibria of an ideal system (n-hexane and n-heptane) and of a non-ideal system (n-hexane-water).

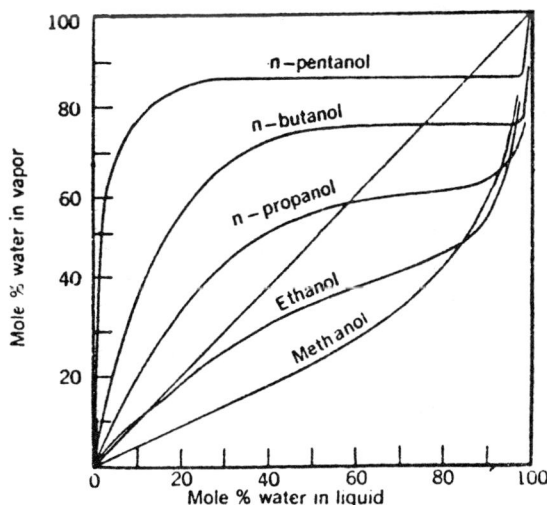

Figure 2. Vapor-liquid equilibria for systems of the lower alcohols and water. Compositions are in mole % water.

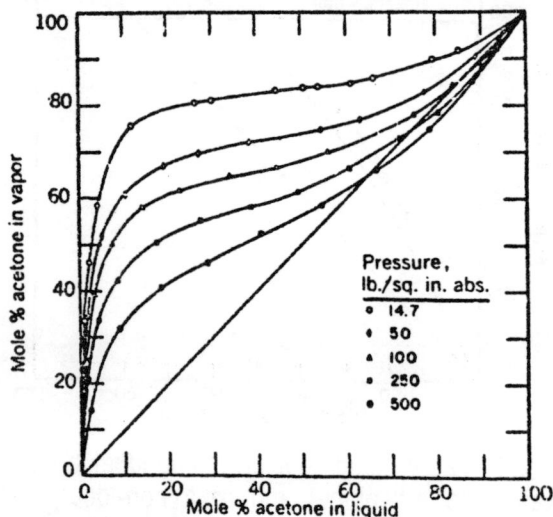

Figure 3. Vapor-liquid equilibria for the system acetone-water as a function of system pressure.
(1 lb./sq. in. abs. = 6.9 Pa.)

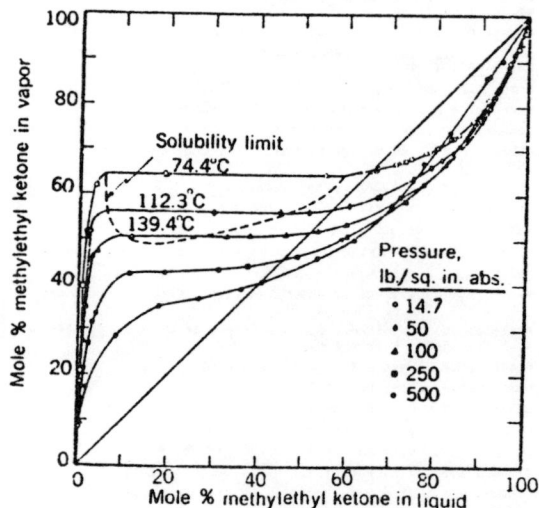

Figure 4. Vapor-liquid equilibria for the system methyl-ethyl ketone water as a function of system pressure.
(1 lb./sq. in. abs. = 6.9 Pa.)

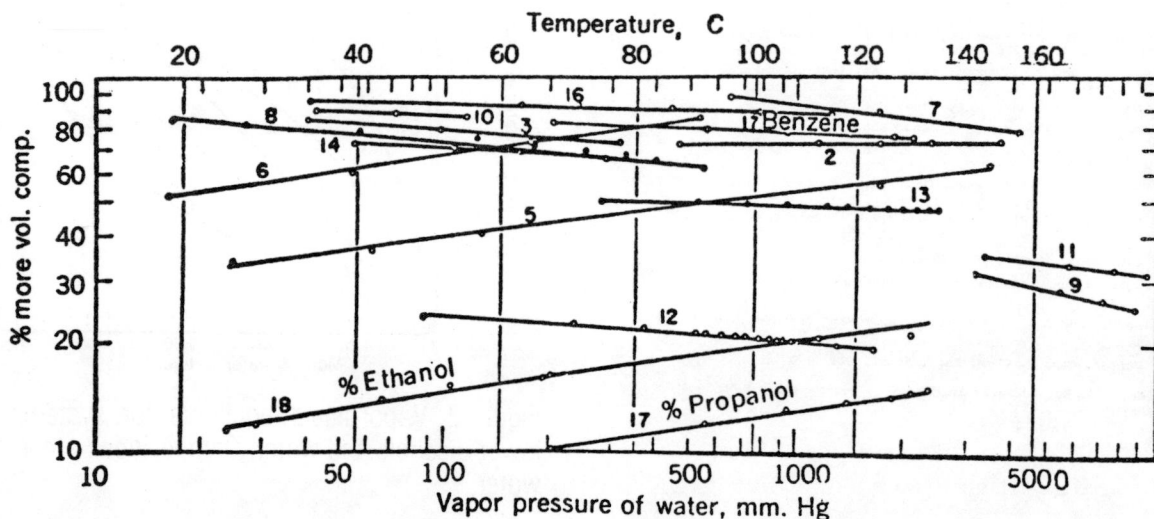

Figure 5. Azeotropic Composition *vs.* Temperature: Vapor composition of the more volatile component *versus* temperature from vapor presure of the reference substance at the same temperature. Composition in the mole or weight percent are as shown in Table 1; systems 4 and 15 fall under 16. For ternary systems, lines are given for two of the components, *i.e.*, benzene and propanol.

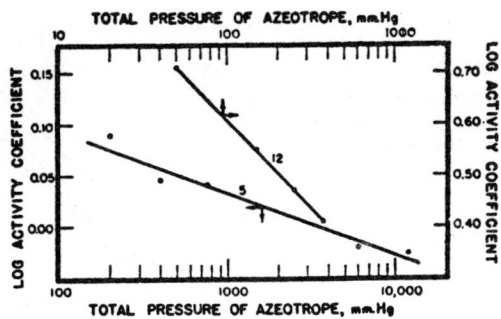

Figure 6. Activity of Coefficient of More Volatile Components *vs.* Total Pressure on System.

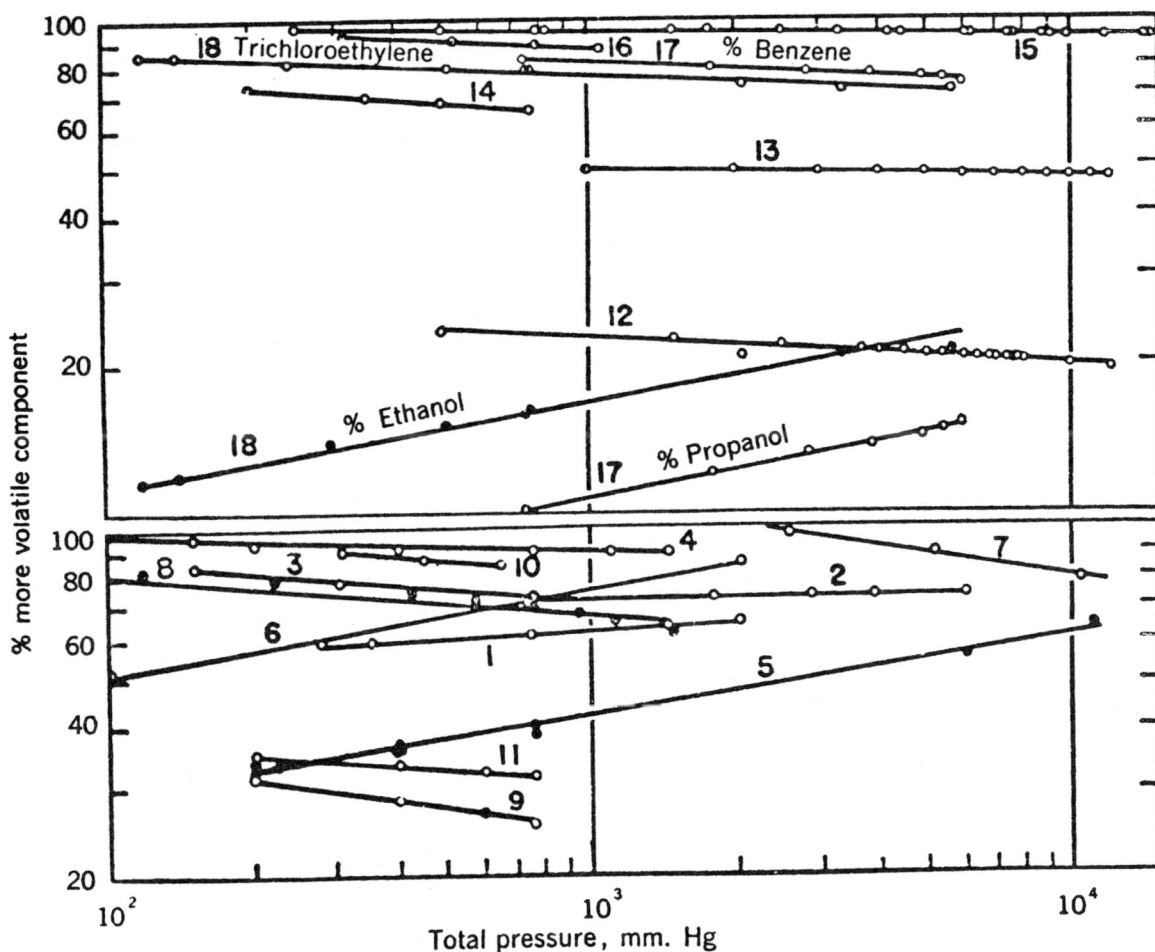

Figure 7. Vapor composition of the more volatile component versus the total pressure on the system. Composition in mole or weight percent are as shown in Table 1. For ternary systems, lines are given for two of the components.

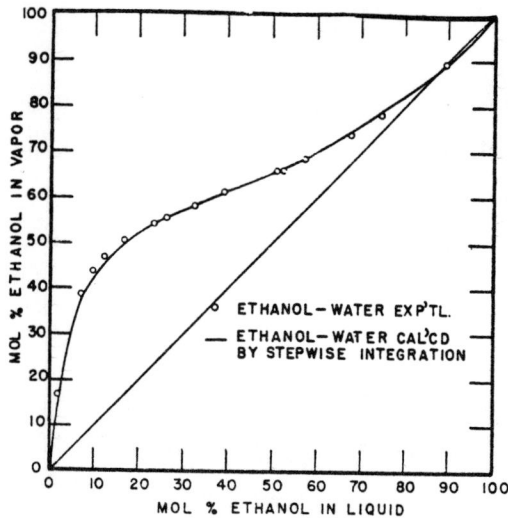

Figure 8. Plot of Mole Per Cent Ethanol in Vapor *vs.* Mole Per Cent Ethanol in Liquid.

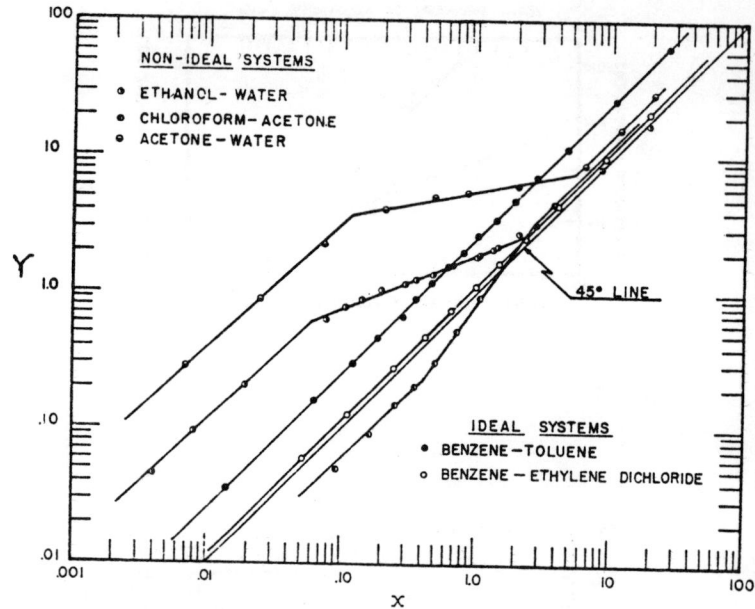

Figure 9. Plot of Y vs. X on Logarithmic Paper for Ideal and Nonideal Binary Systems.

Table 1. Systems plotted Figures 5, 6, and 7.

No.	System	Pressure range, kPa[a]	Composition range[b]	Temperature range, °C
	Homogeneous minimum boiling binary azeotropes			
1	SO_2-*n*-butane	46.6–268.6	59.5–64.4 mol %	−35 to +3
2	*n*-propyl alcohol–water	98.6–790.6	71.7–73.3 wt %	87 to 151
3	acetonitrile–water	20–101.3	72.5–84.5 mol %	34 to 77
4	ethyl alcohol–water	13.3–193.3	89.0–99.6 mol %	34.2 to 95.3
5	methyl alcohol–benzene	26.6–1466.6	33.1–63 wt %	26 to 149
6	methyl alcohol–methyl ethyl ketone	13.3–272	52–85 wt %	18.4 to 92.3
7	acetone–water	342.6–1376	78.0–96.5 mol %	95.8 to 155.8
8	ethyl acetate–ethyl alcohol	3.3–196.73	61.13–87.15 wt %	−1.4 to 91.4
9	β-picoline–phenol	26.6–101.3	25.2–31.5 mol %	146.0 to 187.0
10	chloroform–ethyl alcohol	40.93–86.93	85–90 mol %	35 to 55
11	γ-picoline–phenol	26.6–101.3	31.5–35.0 mol %	147.5 to 190.5
	Homogeneous maximum boiling binary azeotropes			
12	HCl–water	6.6–161.3	19.3–23.5 wt %	48.7 to 123
13	HBr–water	13.3–160	47.03–49.80 wt %	74.12 to 137.34
14	MEK–water	26.6–101.3	65.4–72.2 mol %	39.9 to 73.3
	Heterogeneous minimum boiling binary azeotropes			
15	ethyl acetate–water	3.3–192.173	3.6–9.94 wt %	−1.8 to +89.1
16	ethyl ether–water	4226.6–14133.3	87.5–92.5 wt %	62.5 to 114
	Heterogeneous minimum boiling ternary azeotropes			
17	benzene–propyl alcohol–H$_2$O	98.6–790.6	72.7–82.3 wt % benzene	
			10.1–15.0 wt % *n*-propyl alcohol	67 to 135
18	trichloroethylene–ethyl alcohol–water	15.73–754.6	11.5–21.2 wt % ethyl alcohol	
			70.5–85.1 wt % trichloroethylene	25.1 to 131

[a] In the figures the pressure scale for systems 12, 13, and 15 should be multiplied by 10^{-1}, for system 16 by 10^2. To convert kPa to mm Hg, multiply by 7.50.

[b] Compositions plotted are of the first, more volatile component given except for ternary systems and are either wt % or mol % as indicated under composition range. In Figure 7 the composition scale for system 15 should be multiplied by 10^{-1}.

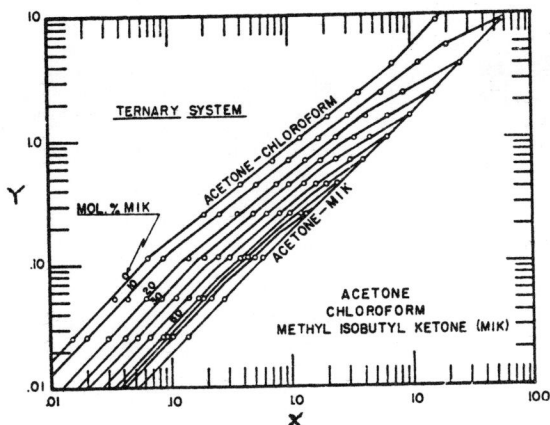

Figure 10. Plot of *Y* vs. *X* on Logarithmic Paper for Ternary System Acetone-Chloroform-Methyl Isolbutyl Ketone.

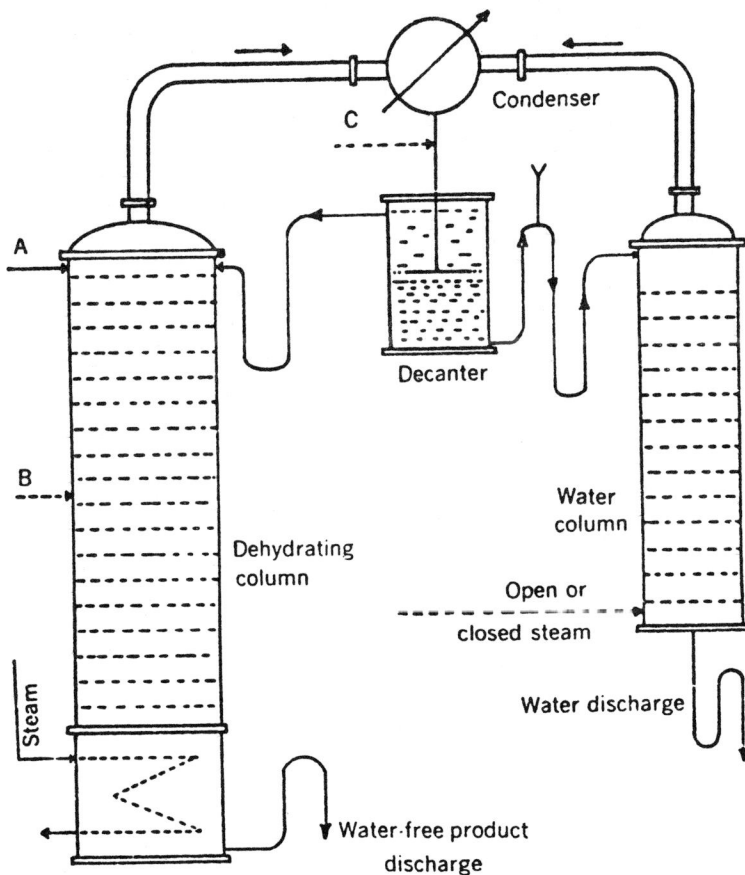

Figure 11. General flowsheet for azeotropic distillation. For water removal, aqueous feed is through line A or B. Water column may have open steam if water is discharged, or closed steam if other liquid is discharged. If it is necessary to separate the condensate into two phases, water is supplied through line C; and in some cases, an extractor may replace the decanter.

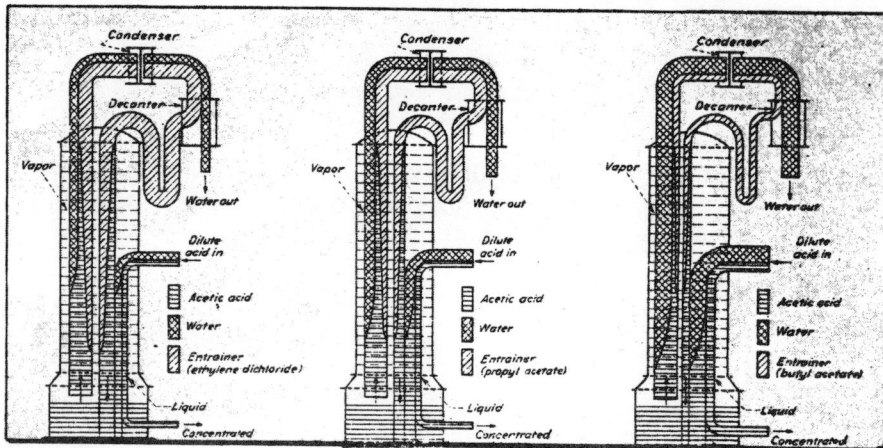

Figure 12. Comparison of the operation of the same column using entrainers with different boiling points in a liquid feed of 33-1/3 mole percent acetic acid. Widths of bands of vapor and liquid are approximately proportional to the number of moles of each component present at any point.

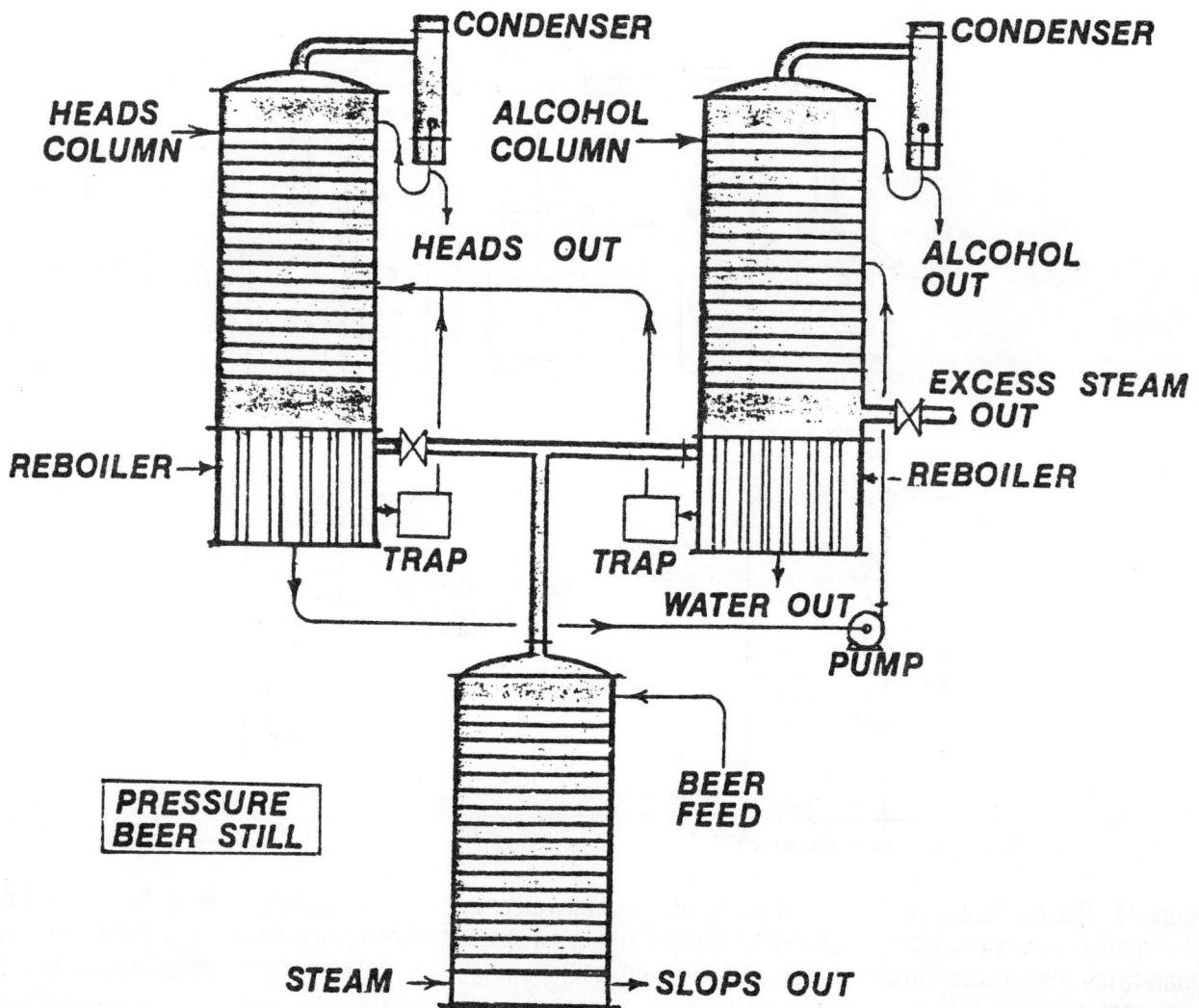

Figure 13. *Vapor Reheat Process.* Vapors from Pressure Beer Still condense in Reboilers of other two columns, condensate goes from traps to mid-section of Heads Column which removes heads and passes alcohol-water alcohol-water mixture to Alcohol Column for usual separation.

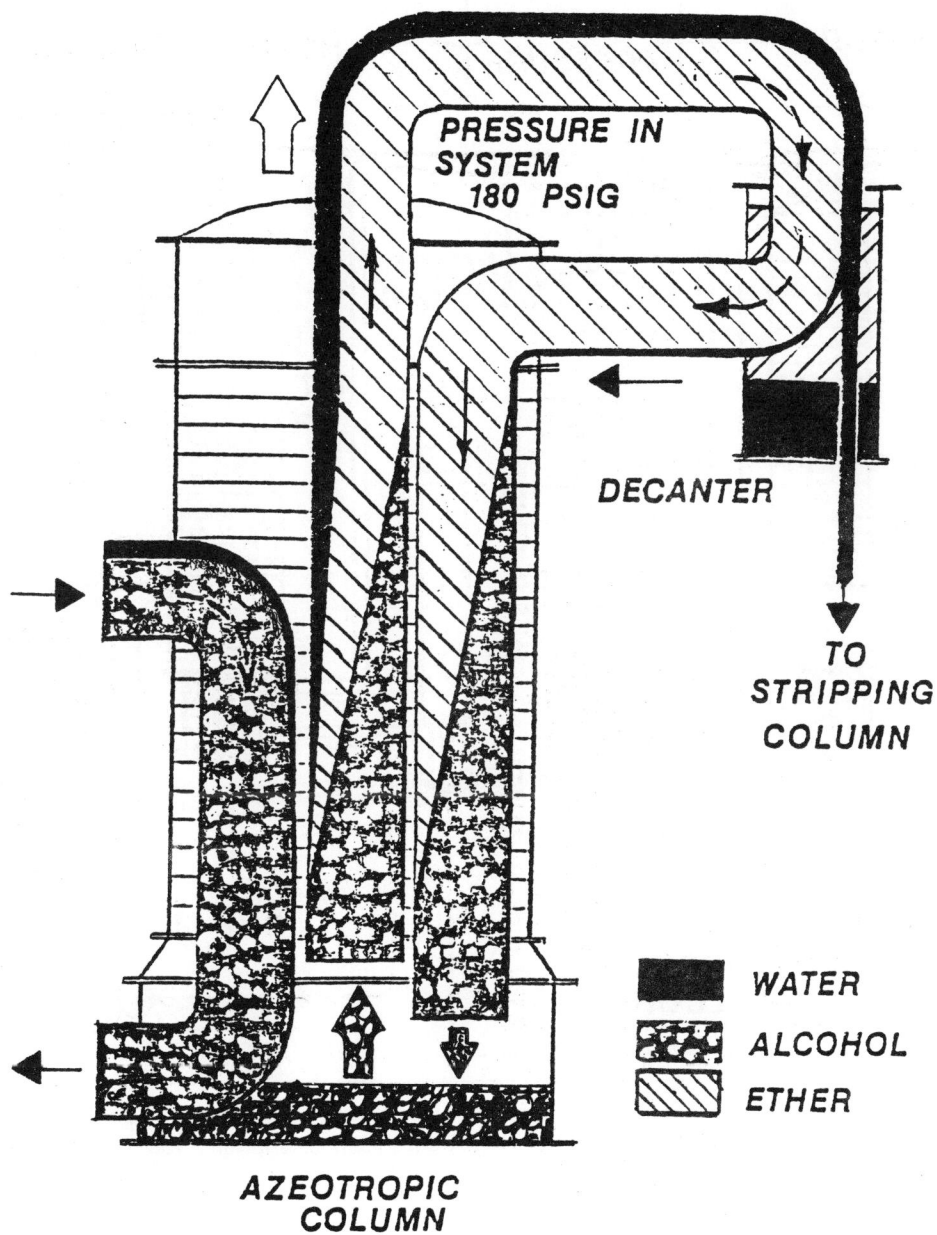

Figure 14. *Diagram of Dehydrating Column (Pressure) using Ether as Entrainer.* Widths of Bands represent approximate molar and heat quantities flowing at each point. Vapor quantitites in middle band, liquid quantities in two side bands.

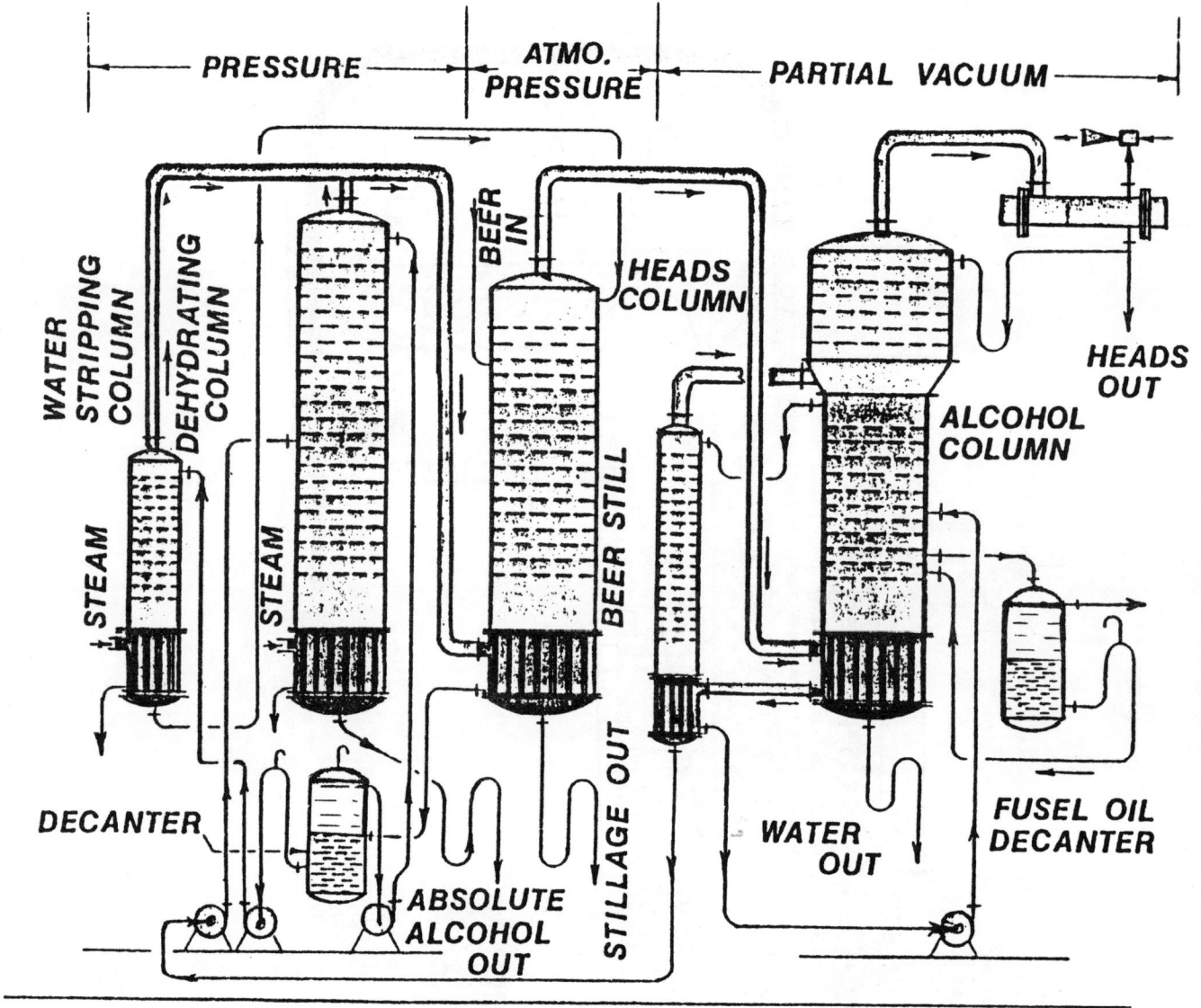

Figure 15. *Production of Absolute Alcohol from Beer.* Low Heat Cost Process, Capacity 150 million liters per year.

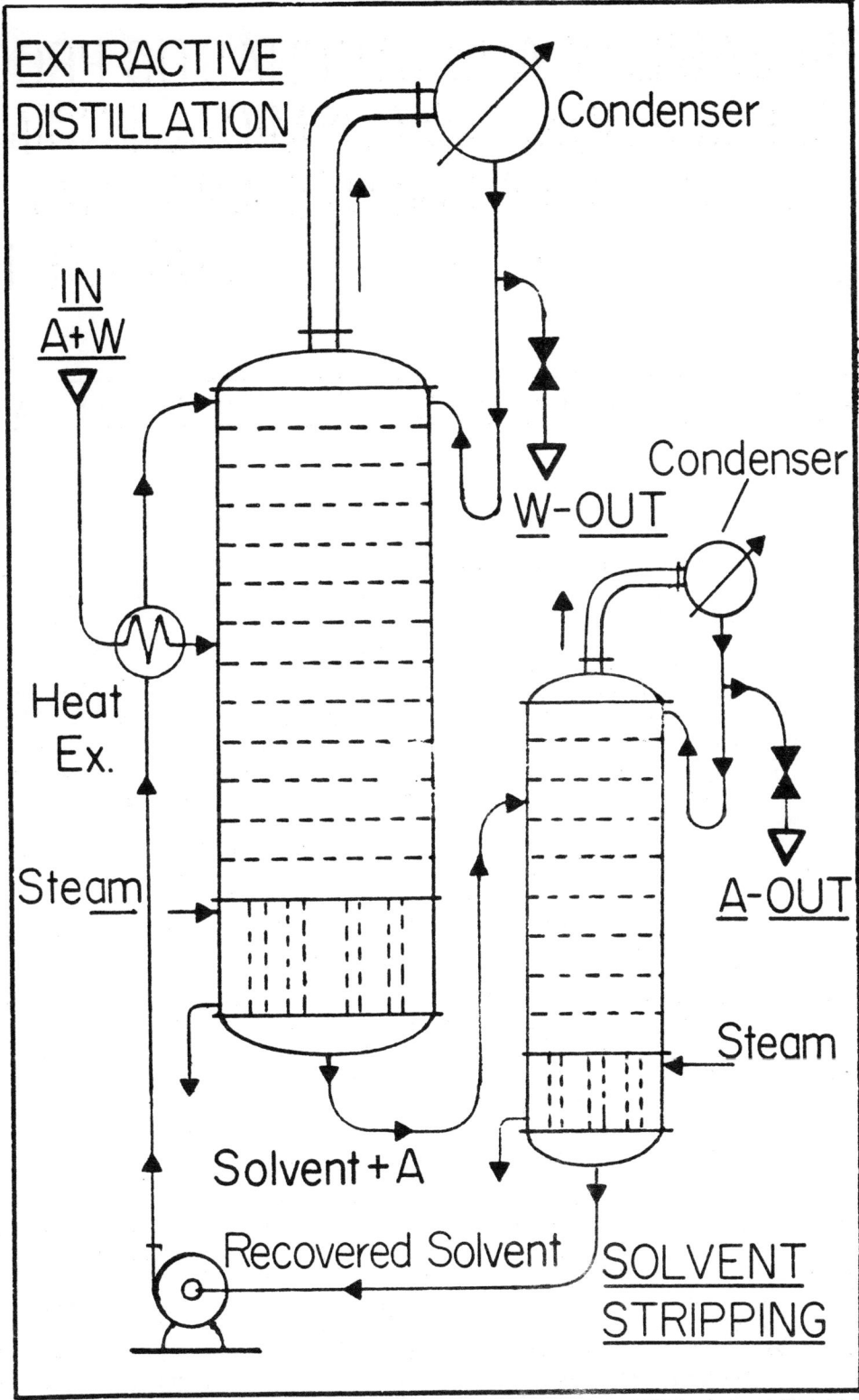

Figure 16. *General Flowsheet for Extractive Distillation.* A Solution of A + W is fed to the side of the Extractive Distillation column and a Solvent for A enters its top plate and decends, extracting A from Vapors. Vapors of W, free of A, are condensed, sometimes there is a slight relfux Solvent plus A leaves bottom; and A is fractionated overhead in Solvent Stripping Column. Recovered Solvent goes out base, through Heat Exchanger and back to Extractive Distillation.

PROBABILITY AND STOCHASTIC PROCESSES APPLICATIONS TO CHEMICAL ENGINEERING PAST ACHIEVEMENTS AND CHALLENGES FOR THE FUTURE

Reuel Shinnar ■ Department of Chemical Engineering, The City College of New York, New York, NY 10031

I feel both honored and somewhat hesitant to give a review in a session called, "Applied Mathematics." While I have had the advantage of having worked in many fields, I have never considered myself an applied mathematician. But I have expertise in one area in which a review article may be useful. Just as I have studied in several languages, I am also multilingual in mathematics. While my knowledge may lack depth, I am fluent and familiar in most of the mathematical techniques used by the chemical engineer. When one speaks several languages one notices that each has its pecularities. While in modern languages one can discuss almost anything in any language, each has areas to which it is specially fitted and others for which it is less adequate. For example, modern Hebrew lacks profane language for daily use; even English cannot compare with Yiddish or Arabic in that area. But both have great advantages compensating for that rather irrelevant deficiency. Now, in mathematical languages, such advantages and disadvantages are much stronger and more significant. My subject will be the one mathematical language for which I have always had a special love, namely, probability. Its value has, as yet, not been fully recognized by the chemical engineer but for me it has been more than a hobby. It has greatly influenced my thinking about my research and design problems. In the following I discuss some unique advantages that it provides for a wide class of chemical engineering problems.

Advantages of Probability Method for Chemical Engineering Problems.

Let me first define what I mean when I say that a mathematical language is suitable or has special advantage for a specific problem. Then we will argue that probability is suitable for problems specific to the chemical engineer and when considering the suitability of a language, there are several properties of which one might think:

A) It is essential to the problem in the sense that it is the only tool available to solve the problem.

B) While there might be other languages available, it is easier to formulate the problem in a specific language.

C) A problem might be formulated in several mathematical languages but in one language it might be easier to grasp the properties of the solution without actually solving the problem. This is especially important in chemical reactor modelling. Good modelling is an iterative procedure, (1). We want a model complex enough to embody the proper physics but not so complex that it is too difficult to solve, to measure the model parameters or to

understand the solution. In formulating a model, therefore, it is important to be able to grasp immediately
whether the formulation has properties
that describe the known physics.
D) Efficient Algorithms. Algorithms
may be more efficient in one language
than in another. An example would be
Monte Carlo methods which are quite
efficient for solving certain difficult
nonlinear problems. The subject of
efficient algorithms will be the only
member of this list of potential advantages that I will not illustrate in
detail because I do not consider myself
an expert in this field and have little
experience in developing algorithms.
For me, the concept of an efficient
algorithm is a good graduate student.
E) Store of Solved Problems. A very
import part of the usefulness of a
spoken language to a specific field is
the size and availability of literature
in that language. The same applies to
mathematical languages. Is there a
library of solutions for similar problems which we can suggest approaches
for how to solve our specific problems?
For certain types of problems in fields
outside of chemical engineering, probability is the accepted language and
offers a rich cache of solved problems.
F) Generation of New Ideas. Very
often the formulation of a problem in a
specific language leads in a natural
way to an approach for solving the
problem. Once we have solved it, we
might translate the solution into other
languages, but the language in which we
formulated the problem was itself
instrumental in generating these ideas.
An example of this is Wei's and
Prater's work on the structure of chemical reactions which resulted from
their interest in linear algebra.

All these advantages can be found
in the application of probability to
certain classes of chemical engineering
problems. The purpose of this paper is
to demonstrate this by specific examples.

Probability, however, is not a
general purpose language nor even a
prime language for the chemical engineer. I would only recommend its use
for problems where it has a specific
advantage. This is especially so,
since the average chemical engineer

(and even a majority of our faculty),
is not familiar with it. This is very
regrettable (because they lack an important tool), but nonetheless a fact.
Very often the only solution is to
translate one's results into a more
familiar language or to explain the results in a way that the chemical engineer can understand them without requiring a thorough knowledge of probability theory.

The main thrust of the following
review is to demonstrate each of the
above points with a few simple, suitable examples that can be understood
without a detailed knowledge of probability theory. The emphasis will be
on explaining the ideas. No attempt
is intended to cover the wide literature in the field. The reader will
hopefully excuse the fact that the majority of the examples given below
come from my own work.

Areas of Chemical Engineering for
which Probability is Essential.

Actually, it is strange that most
chemical engineering departments do
not include at least some fundamental
probability theory in their basic undergraduate or graduate programs. For
the practicing engineer, especially
those going into production or management, probability theory is an essential tool (see Table 1). In a large
number of industries, such as polymers,
textile fibers, pigments, metals, the
control of the specific processes,
etc., is in large part, based on
sampled data quality control (2,3),
thus an engineer must understand the
problems of sampling and measurement
errors. He also must be able to design
experiments based on statistical techniques (4,5). He must deal with
hazards and reliability (6).

Table 1 Problems for Which Probability
         is Essential
1) Statistical Quality Control
    Sampling
    Measurement errors
    Design of experiments
    Reliability analysis of design
2) Unsteady processes
    Turbulence
3) Decisions under uncertainty
    Operational research
    Risk analysis

In research there are several areas where probability is essential, such as in turbulence, transfer processes to a turbulent film, and others (7,8). Last but not least, we note that a large fraction of our graduates end up neither in design nor in research, but in management. If they want to understand modern management theory, probability is an essential tool (9); it is the language spoken in operational research, risk analysis, etc. Teaching the student probability from the beginning will make the graduate more versatile. But in my review I will not focus on these obvious areas, mainly because our profession has contributed little to either statistical quality control of to modern management theory. Instead, I will focus on applications of probability in chemical engineering research. To show the power of these methods, I choose examples in which the need for using probability is not totally obvious. Let me start with item B from the above list.

Ease of Formulation

There is a large class of problems that has a dual nature. The system itself is completely deterministic, but can be described in a probabilistic way by looking at single particles or molecules. If we look at a single particle we cannot predict its exact future; we can only represent it in probabilistic terms. However, when the number of particles or molecules is extremely large, the (macroscopic) overall behavior of the system is completely deterministic. We have here two completely complementary descriptions of the same systems, similar to Eulerian and Lagrangian formulation of fluid dynamic problems. Which approach one should choose strongly depends on the case in point, but sometimes using both approaches results in synergistic information and additional insights. Probability is useful because it is easier to write the equations, to formulate the boundary conditions, to grasp the solutions and to provide insights into the properties of the system. A list of such research areas is given in Table 2. Some suitable reviews and examples of such problems are ref. (10) to (18).

I will limit myself here to one example from polymerization. Consider a two phase polymerization reactor in which the monomer is dispersed in a large number of droplets which are present in a water phase, and in which the catalyst generates free radicals in the water phase as well. Emulsion, suspension and dispersion polymerization are examples of such systems.

Table 2   Deterministic Problems for which Probability is Sometimes Preferable Language

   Particulate Systems
   Polymerization
   Residence Time Distributions
   First Order Reaction Systems
   Diffusion

Total System   →   Deterministic
Behavior of Single Particle Probabilistic

We will assume that the polymerization itself is a simple free radical polymerization where growth occurs by addition of monomer to a free radical and polymer particles are formed by termination of two free radicals with each other. If the system is homogeneous we can describe the evolution of the molecular weight distribution by some simple population balance (19,20).

$$K \xrightarrow{k_o} R_o$$

$$R_j + M \xrightarrow{k_p} R_{j+1} \qquad j = 0.1\ldots\infty$$

$$R_j + R_k \xrightarrow{k_t} P_{j+k} \qquad j,k, = 0.1\ldots\infty \qquad (1)$$

$$k_p[R_j+1][M] + k_p[R_j][M] - k_t[R_j][\sum_o^\infty R_k] = 0$$

where K is the catalyst, R_j a free radical of size j and P_j a polymer molecule of size j. The population balance can be solved to give the molecular weight distribution

$$P(n) = \left(\frac{2n}{<n>} + 1\right) e^{-2n<n>} \qquad (2)$$

We can derive the mass balances of eq. 1 from the underlying probability equations (13), but this offers no advantage here. If we consider the two phase case the simple behavior described by eq.(1) does not necessarily hold anymore. We now have a large number of droplets; a free radical growing in one droplet cannot terminate with a radical in another droplet. For (1) to hold, there must be a large number of free radicals in each droplet. What happens if the number of growing radicals is small? The simplest case is the well-known limiting case of idealized emulsion polymerization (21). Here we assume that the termination rate is so large compared to the growth rate of a free radical (the rate at which it adds monomer) that if a free radical enters a small particle and finds another free radical it immediately terminates. If it finds none it adds monomer and grows until a second free radical enters. This is a simple and, obviously, deterministic system when the total number of growing radicals is very large. But, it is difficult to even write out the deterministic population balance. On the other hand, if we look at a single free radical and try to compute its probability to grow to a certain size, the problem becomes simple and almost obvious. The probability of a single free radical growing to a certain size depends on the time it can spend alone in a particle before the next free radical enters. Therefore, the molecular weight distribution is proportional to the distribution of inter-arrival times of free radicals into a single particle. If the free radicals are generated at constant rate then a simple probability argument shows that the distribution of these times must be Poisson. It follows that the molecular weight distribution is also Poisson and is of the form

$$P(n) = e^{-n/<n>} \qquad (3)$$

where <n> is the number average molecular weight.

We note that eq.(3) is completely different from eq. (2). Eq. (3) is an ideal limiting case, though it describes some practical cases quite well. It applies when the number of growing radicals per particle is either zero or one. What happens when the average number of growing radicals per particle is two? In that case, we cannot even apply the population balance of eq.(1). Eq. (1), as all deterministic population balances of this type, assumes that if a growing free radical terminates, the chance that the other radical with which it reacts is of size j is simply the fraction of all radicals which are of size j. Intuitively, one realizes that this is not likely if

there are only two radicals per particle. This fact implies that termination rates are rather large compared with the rates at which particles enter the system: if there are two growing free radicals in one particle and one of them is large, then the likelihood that the other is also large is small, or at least much smaller than the independent probability of a single growing free radical to be large. The fact that one radical is large implies that it was the only radical present for a significant time. Therefore, most probably, the second free radical arrived fairly recently. The statistician calls this interaction, which means that the probability of the second particle to be of size j depends on the size and number of the other particles in the system. The population balance in eq. (1) implicitly assumes that this is not the case. When we write down a simple deterministic population balance we may not even notice that we are assuming this. If we derive eq. (1) by the more complex method of averaging the probability histories of single radicals, we learn which assumptions we must invoke. Using a probability formulation we can solve the problem for an arbitrary number of growing free radicals per particle (13). This method is cumbersome, but, at present, it is the only one that can solve problems of this type. The solution shows that if the number of growing free radicals, n, exceeds 20, the molecular weight distribution is almost indistinguishable from the case of n→∞. This gives us insight into the importance of interaction for a given system (see for example (13,15,16,17)).

The preceding problem could probably be solved by deterministic methods if one knows what the solution looks like. This is because, in essence, it is a deterministic problem. However, using a probability approach it is much easier to formulate the problem and to determine initial and boundary conditions. It also shows the importance of taking into account interactions which are often overlooked in deriving simple mass balances. Some areas of chemical enineering in which interactions are important are given in

Table 3. There is a large number of problems in the areas outlined in Tables 2 and 3 for which a probability approach can be useful.

Table 3   Population Balances with
          Interactions
                  Polymerization
                  Coalescence
                  Flotation
                  Crystallization

Grasping the Solution of a Problem Without Actually Solving It

In this section, I want to bring two examples from chemical reactor design involving the use of residence time distribution functions. Consider a steady state reactor such as the one described in Fig. 1. If we introduce into the inlet a "suitable" tracer with a linear response, we can write that

$$C_p = f(t) * C_F(t)$$

where $f(t)$ is the transfer function of the system and the * indicates the convolution product. The function $f(t)$ has a dual interpretation (22,23,24). It can be interpreted simply as a linear transfer function or it can be given a probabalistic meaning. $f(t)$ is the probability that a single particle entering the system at time zero will leave the system between the times t and $t+\Delta t$. Therefore $f(t)$ is called the residence time density function of the system.

RESIDENCE TIME DISTRIBUTION

$C_o(t)$ → □ → $C_p(t)$

$C_p(t) = f(t) * C_o(t)$ **TRANSFER FUNCTION**

$f(t) \, \Delta t$ **PROBABILITY OF PARTICLE TO LEAVE REACTOR BETWEEN TIME t AND t+Δt**

Figure 1

What advantage accrues by giving $f(t)$ a probabalistic interpretation? If we want to compare the response of the system to a specific preconceived model (such as three stirred tanks in series) there is really no advantage; in either case one computes the expected form of $f(t)$ and compares it with the actual response. However, the probability approach has several conceptual advant-

ages, which we illustrate with two examples. In the first example, we can transform $f(t)$ into a different type of probability distribution, a so-called conditional probability distribution. This distribution represents the answer to the question, what is the probability that a particle leaves the system between time t and $t+\Delta t$, provided that at time t it is still in the system? We designate this probability $\Lambda(t)$ and can show (23) that

$\Lambda(t)$ is called the escape probability or the intensity function and has one advantage: We can easily grasp its physical significance.

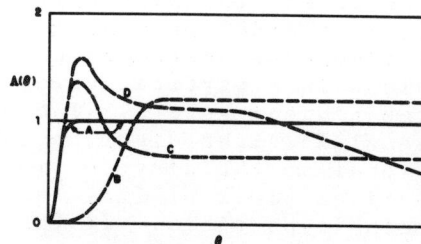

INTENSITY FUNCTION FOR IMPERFECTLY MIXED STIRRED TANK REACTORS
A. SHORT DELAY BETWEEN INLET AND OUTLET
B. DELAY BETWEEN INLET AND OUTLET DUE TO INSUFFICIENT STIRRING
C. BYPASS BETWEEN INLET AND OUTLET
D. BYPASS BETWEEN INLET AND OUTLET AND STAGNENT REGIONS

Figure 2

All positions in a stirred tank are, in a sense, equal. Therefore, $\Lambda(t)$ should be a constant, that is, equal to $1/\tau$ where τ is the average residence time for the reactor (see curve A, Fig. 2). In practice, one knows that it is impossible to achieve an ideal stirred tank; that is, it must take a non-zero time for an entering particle to reach the exit. $\Lambda(t)$ should therefore look like curve B in Fig. 2. Consider a viscous system in a tall vessel with feed at the bottom and outlet at the top where the stirring is not very intense. We expect that a latency period immediately following a particle's introduction into the system when that particle has practically zero probability of escaping is large, however, after this latency period, $\Lambda(t)$ should become constant as in a stirred tank. Consider now a tank in which there is maldistribution. Part of the feed bypasses the tank and escapes before it becomes mixed. That might happen if the inlet and outlet are too close together and the mixing is not sufficiently intense. During some initial time, immediately following its introduction, the probability of a particle leaving the system is much larger than later. If a particle has been in for a long time, it is clear that it was not in the bypass and, therefore, its probability to exit given that it is still in the vessel after a long time is less than its initial exit probability. Thus, we expect $\Lambda(t)$ to exhibit a maximum and look like curve D. If we measure an actual tracer response from a real reactor vessel which does not perform as expected, we can use arguments such as these for form of $\Lambda(t)$ in order to diagnose what might be the source of the malfunction. We note that we were able to construct the form of $\Lambda(t)$ for these various cases without solving any equations. This illustrates what we mean by being able to grasp the properties of a solution directly from the physics of the system.

Use of $\Lambda(t)$ has also another significant advantage. The form of probability distributions such as $f(t)$ is severely constrained by the fact that they must be normalized, i.e., that $\int_0^\infty f(t)dt=1.0$. Conditional probability distributions have much less rigorous constraints on their form and therefore are much more suitable in evaluating the nature of differences in the response. The example also illustrates the advantage of having available a library of similar problems in a given language. The idea of using $\Lambda(t)$ came from other problems in which a probability approach is commonly used.

Network of Stirred Tank Reactors
Optimal Configuration for first order reactions

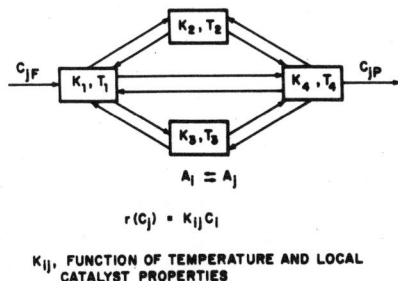

$A_i \rightleftharpoons A_j$

$r(C_j) = K_{ij}C_i$

K_{ij}, FUNCTION OF TEMPERATURE AND LOCAL
CATALYST PROPERTIES

Figure 3

Our second example will deal with a well known optimal reactor design problem. Consider an arbitrary network of stirred tanks as given in Fig. 3. A complex first order reaction occurs in these tanks. The tanks are each ideally stirred and each maintained isothermal at any prescribed temperature which is independent of the reactions occurring in it. The temperature of each tank may be different and each tank may contain a different catalyst. The reactions involve a set of N species $A_i(i=1,2,3..N)$ which even are related by a set of reactions.

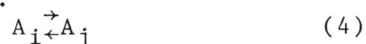

$$A_i \rightleftharpoons A_j \qquad (4)$$

These reaction rates are linear in the sense that for each tank we can write a reaction rate for the formation or disappearances of species A_j in tank n

$$r(C_j) = \sum_{i=1}^{N} k_{ij}(n)C_i - \sum_{i=1}^{N} k_{ji}(n)C_j \qquad (5)$$

C_j and C_i are the local concentration of species A_i and A_j and k_{ij} is the matrix of linear reaction rate constants. k_{ij} may, however, be different in different tanks. We now ask what is an optimum configuration for such a reactor? We can deduce some important information about the nature of the optimal reactor system by using a probability approach and without solving a single equation (25).

We can describe the system in question in a different manner which leads to identical equations but with different physical meanings. Consider a system of particles each of which can have N different state which we designate $A_i(i=1,2..N)$. They can change these states spontaneously and the transition probability is given by a matrix k_{ij}, where $k_{ij}t$ is the probability that a particle in state i to change to state j in the interval Δt. The probability that particle is in state i at time t is defined by $p_i(t)$. If the total number of particles is very large the probability of a particle to be in state i becomes equal to the fraction of particles in state i, which is proportional to the concentration, c_i, of particles in state i. This results in what the probabilist calls a Markov process. We may employ the well-known mathematical tools to

describe the development of the probabilities $(p_i(t): i=1,..,N)$ with time.

In a conventional chemical engineering manner, one would solve the
system in Fig. 3 by writing up a set of
mass balances for each compound in each
tank. This will result in a set of
linear algebraic equations, which can
be solved for any specific network,
given the sizes of the tanks, the flows
and the values of the matrices $k_{ij}(n)$.

However, while the set of equations is linear in c_{ij}, it is not linear in terms of the variables we look
for in terms of optimization. These
variables include the volumes of the
tanks, the interconnecting flows and
the sets $k_{ij}(n)$ which we can change by
changing the temperature of individual
tanks. It is therefore a difficult
optimization problem. What do we gain
by reinterpreting this problem as one
invoking a Markov process? If the system is given and I seek an actual solution for the product distribution
$(c_j: j=1,..,N)$ at the reactor outlet, I
gain absolutely nothing; I will end up
with exactly the same set of equations,
but it will only be more difficult to
derive them. However, the formulation
in terms of a Markov process allows me
to grasp some important properties of
the solution without actually solving
the resulting equations.

Let me do a gedanken experiment.
I will stand at the outlet of the reactor and interview each particle assuming that it has a perfect memory. I
will note down exactly not only the
time the particle spent in each vessel
but also in which vessel it was at
every instant. This I call the particle's history. I will now sort them
into bins such that all particles in
each bin have histories. I further
assume that the number of particles is
so large that I still have a large number of particles in each bin (see
Fig. 4).

Figure 4

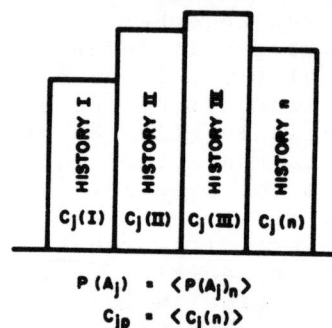

$$P(A_j) = \langle P(A_j)_n \rangle$$
$$c_{jp} = \langle c_j(n) \rangle$$

Sorting the particles of the
reactor in Fig. 3 into bins
according to the total detailed
history of the particles.

First, I note an important fact.
If I look at the history characterizing
the particles in an individual bin I
note that a particle could have realized exactly the same history in a plug
flow reactor divided into sections such
that in each section the set of
$k_{ij},..,j=1,2..3,N)$ is constant (see
Fig. 5). The length of each section is
not uniform because it corresponds exactly to the time that the particles
in this bin spent in a specific vessel.
If they returned to a vessel several
times, the plug flow reactor will have
appropriate sections for each visit.

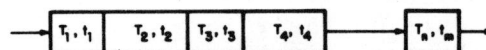

EQUIVALENT REPRESENTATION OF A PARTICLE HISTORY IN A
NONISOTHERMAL REACTOR. SECTIONALLY UNIFORM PLUG FLOW
REACTOR.
T_i – TEMPERATURE OF EACH SECTION
t_i – RESIDENCE TIME OF EACH SECTION

Figure 5

If we know the complete history of
a particle at the exit, we can compute
its conditional probability of being in
each of the states j given that it en-
tered in state i. The probability distribution p_j (given history) for the
bin is also proportional to the concentration in the bin. In fact, if we
normalize the inlet concentration, we
can show that $c_j=p_j$. We could have
also computed $p_j(bin)=c_j(bin)$ by solving the equations for the plug flow
reactor in Fig. 4. If we now want the
outlet concentration of the total

reactor all we have to do is to average over all of the bins, i.e.,

$$c_{jp} = <C_j \text{ bins}> = <p_j(\text{bins})> \qquad (6)$$

If we use this procedure to compute c_{jp}, we could set up a probability machinery to generate all of the histories of the bins. But we are not interested in that. All we wanted to show is that we derived the final solution for $<C_p>$ by averaging over the individual bins or histories.

Assume now that somebody has designed such a complex network of stirred tank reactors and claims it has some optimal properties in the sense that one of the C_j has either a maximum or a minimum yield at the outlet. If this is true then all we need do is to recall a simple algebraic result. In any set of numbers there is a single number of the set which is either larger or smaller than the average, unless all the numbers of the set are equal. Since the c_{pj} of this complex network could be derived by averaging over a set of bins, there must be a bin for which the values of C_j is bigger or smaller than $<C_j>$ for the network. This means that I can always design a plug flow reactor with a temperature profile (or different catalyst sections) which is as good or better than any network of stirred tanks (25).

This is only correct if the reactions are first order (or pseudo first order) and if the temperatures in each reactor are not affected by the reaction itself. For example, had the reactors been adiabatic, knowing the probability history of a single particle would not have allowed us to draw any conclusions about its state. The reaction rates or transition probability would have been affected by the state of his neighbors and simple averaging would have been meaningless.

Using simple probability concepts allowed me to derive an important result in reactor design without solving a single equation. All I had to realize was that in solving the equations I really average over independent histories. Whenever I do this, a plug flow reactor (corresponding to a single uniform history) must be optimum. There are a considerable number of problems

in reaction engineering for which probability offers a framework for organizing our thoughts similar to the one just discussed (1,26).

In no way do I claim that this is a specific property of probability. In many chemical engineering problems one can do the same, using geometry, vector algebra, matrices, etc. For each problem there might be a preferred mathematical language in which it is easiest to grasp the properties of the solution by just looking at either the equations or the formulation of the problem; it pays to be multilingual and to search for that language. All I wanted to show here is that there is a large set of chemical engineering problems for which probability is that natural language.

Generation of New Ideas

I mentioned in the introduction that formulating a problem in a specific language may generate new ideas and approaches not just for solving a specific mathematical problem, but for approaching an engineering problem.

I would like to give an example of this from modern control theory. The most important developments in modern control theory have their roots in work done using a probabilistic or stochastic approach. The main results that come out of this work can be translated to classical control theory, and it is interesting that only after having been translated did they realize wider application.

One can look at the feedback control of a chemical process in two ways. The conventional way is given in Fig. 6a. We want to keep the output Y of a process at some given value in the face of a disturbance n_t. We measure the output variable Y and compare it to a derived value. We then feed the difference to a controller which, using a controller algorithm G_c, computes a control action of a manipulated variable. Controller design is mainly concerned with finding an algorithm G_c, that achieves this for a given process, described by a process model G_p in a way that guarantees the stability of the system. (Stability here is defined as asymptotically stable for small disturbances.)

There is another way to look at
the same problem that has its intellectual roots in the problem solved by
Wiener for antiaircraft gun control
(27). If we want to fire at an aircraft it is clear that we would not aim
at the aircraft because by the time
that the projectile reaches the position aimed at, several seconds have
passed and the aircraft will have moved
far away. Therefore, we must estimate
where the aircraft will be when the projectile arrives. We do this by two
separate estimates, an estimate of the
future flight of the aircraft based on
its previous trajectory and an estimate
of the trajectory of the bullet as a
function of time and aim. A skilled
hunter aiming at a running dear does a
similar calculation intuitively. Since
the aircraft problem is complex, an algorithm is helpful.

We can look at control in the same
way (Fig. 6b) or Ref. (28,29). If we
change a manipulated variable in a process input, the output does not respond
immediately; there are often delays
present. We can look at the process
model as a prediction maneuver, of what
will occur to Y(t) as a result of any
control action U. Moreover, process
disturbances are not totally random
(i.e., they are not "white noise."
They show a correlation with time, and
we can guess, based on the observed
history of Y(t), what its value will be
in the next time interval. Just as in
the antiaircraft problem our guess will
be imperfect. However, it will form a
reasonable basis for a control action.
In practice, what this means is that we
must build a model for the disturbance
n_t that is based on the past knowledge
of the disturbance and precicts its
future behavior. We try to match the
impact of this disturbance at some
future time, t+k, with a control action
now that exactly compensates for this
disturbance.

If we look a little closer we find
that the difference between the two
schemes is less than is apparent at
first. In order to predict the future
impact of the disturbance, the scheme
in 6b uses past values of the disturbance. But since these are not directly

measurable it must extract them first
from past measurements of Y. If the
controller in 6a has an integral control action, it also computes the necessary control action based on past
values of Y.

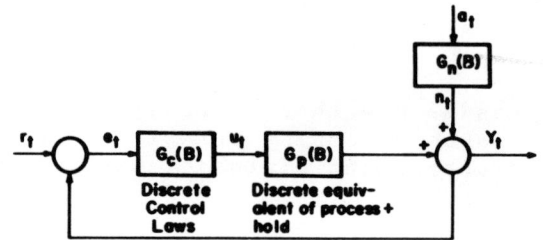

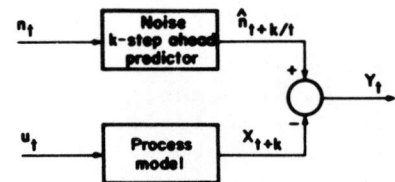

Fig. 6 Control schemes for
        process control.

   6a Conventional control scheme
   6b Control scheme based on
      matching two predictions.

In order to make the predictions
needed in 6b, we need a mathematical
model of n_t. This is normally called
a filter to which white noise is fed.
Once we choose the mathematical form
of this filter we can design a feedback
controller and represent that feedback
controller in purely classical terms
without noting that it was conceived as
a noise predictor.

Let me just give a simple example.
A popular model for a process disturbance is the one formulated by Box and
Jenkins (32) and called an Arima
(0.1.1) disturbance. It has the form

$$n_t = \frac{1-\lambda}{1-B} a_t \qquad (7)$$

where B is the backward operation
($By_t = Y_{t-1}$) and a_t is a white noise

signal.

Consider a first order system with a delay that has the continuous transfer function

$$G_p(s) = \frac{K_c e^{-\theta s}}{1 + \tau s} \qquad (8a)$$

For a sampling time T the discrete transfer function becomes

$$G_p(B) = \frac{\omega_o - \omega_1 B}{1 - \delta B} B^{k+1} \qquad (8b)$$

where ω_o, ω_1 and δ are system parameters which are uniquely determinable of τ, θ, K_c and the sampling time T. Using Fig. 6b will get the following:

$$G_c(B) = \frac{\frac{1}{\omega_o} - (1 - \delta B)}{1 - \frac{\omega_1}{\omega_o} B [1 - \lambda B + (\lambda - 1)B^{k+1}]} \qquad (9)$$

This is really a simple proportional integral controller with a dead time compensator. The gain of this controller is a function of λ, the only free parameter in eq. (7). If $\lambda = 1$, the gain is zero. This is sensible because for $\lambda = 1$ the disturbance is really white noise. If there is no correlation between successive values of n_t, there is no sense in using control. The mathematical details of deriving this controller are of no interest to us here and the reader is referred to the literature (28,32). What does interest us here are some special features of the controller derived from Fig. 6b.

While the controller in 6b always has an equivalent formulation in terms of conventional feedback control, that formulation has some specific unique properties that made a very valuable contribution to conventional feedback controller's design. If we reformulate 6b in a form similar to 6a, we will always end up with a controller with a structure similar to the one given in Fig. 7 (28,30,31). It will contain the equivalent of the process predictor in Fig. 6b. If we observe a deviation in the output variable and compute a required compensating action, it makes sense to wait until this action is fully felt at the outlet. However, we cannot wait because other disturbances may appear while we are waiting. What we can do (and is done in Fig. 7) is to

first correct the measured value of Y by deducting from it the estimated impact of all the control actions that we have made, but, as yet, have not affected on the measured Y. Such a scheme is called either inferential control or internal model control (30,31). It is really a translation of stochastic control theory into a conventional control formulation which was suggested in Ref. (28) and has since then been applied to multivariable control (31).

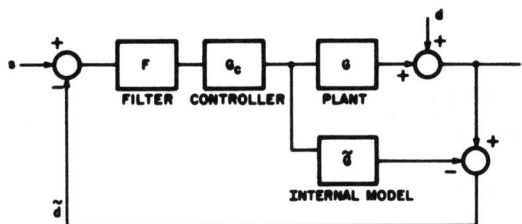

IMC STRUCTURE WITH FILTER.

Fig. 7   Translation of control scheme in Fig. 6b into a conventional control scheme representation (internal model control or inferential control).

For a simple invertible process with dead time Fig. 7 can also be represented by the scheme in Fig. 8, which is a simple dead time compensator with a filter, which is the controller in Eq. (9). Fig. 7 also contains a filter in the feedback circuit. The Box and Jenkins formulation (or the equivalent formulation by Kalman (32)) contains this filter due to the formulation of the noise structure. Here the conventional formulation has an advantage because we can easily see what the purpose of this filter is. The scheme in Fig. 7 has a problem that it shares with all stochastic optimal controllers. The controllers obtained contain the inverse of the process transfer function. Even if the transfer function is not invertible it contains something which is very similar to an inverse (28). This leads to a particular type of stability problem. In terms of conventional stability analysis such controllers are always stable, but very sensitive to G_p. We show this in Fig. 9 where we give stability limits for the

controller given by Eq. 9. Here, the
stability limits are given in terms of
the parameters of G_p, τ and Θ (eq. 8a).
We concern ourselves here only with the
basic ideas; the reader interested in
the details should consult Ref. (28 to
31).

In the noise formulation of Box
and Jenkins λ can vary between -1 and
+1. We note however that for values of
less than 0.5 the controller is very
sensitive to variation in process parameters. This phenomenon is what we call
high model sensitivity.

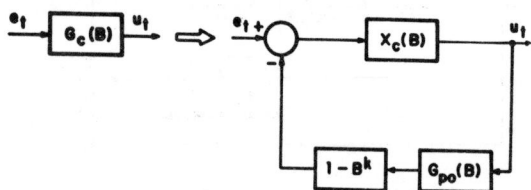

Fig. 8   Alternative structure of
         controllers in Fig. 7 for
         simple process with dead
         time. (Sampled data control)
         Process model without dead
         time G_{po}
         Dead time k sampling
         intervals
         B backwards operator defined
         by $n_{t-1} = Bn_t$

The classical control engineer
stays away from any controller that
contains an inverse of the transfer
function because this cancels poles and
zeros in the characteristic equation.
If we do not know G_p exactly such cancellations are mathematical artifacts
and consequently they give us a false
sense of security. However, such controllers are inherently sensible. Instead of dropping them, all we need to
do is to modify our stability analysis.
We must take into account the fact that
in Fig. 6b the prediction of the impact of a control action also involves
uncertainties, just as the prediction
of the future disturbance. We can take
care of this if we desensitize the system by introducing a filter. Making
the filter strong enough will always
assure stability but will reduce the
quality of the control.

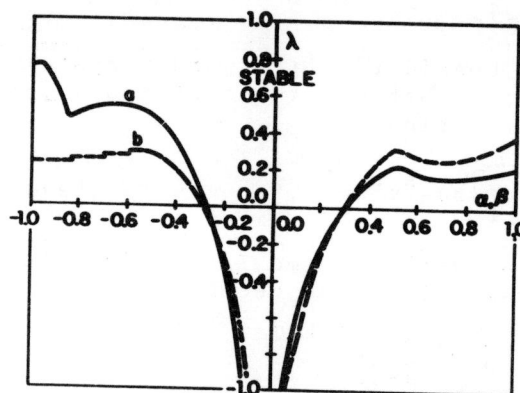

STABILITY LIMITS

$$G_p(s) = \frac{e^{-\theta s}}{1+\tau s}$$

a) $\tau = 1 \pm \alpha$

b) $\theta = 0.5(1 \pm \beta)$

Fig. 9   Stability limits for a Box
         and Jenkins sampled data (eq. 9)
         operating on a continuous pro-
         cess with the transfer function
         $G_p(s) = e^{-0.5s}/1+s$. Sampling time
         interval 0.25 time units.
         Stability limits are given in
         terms of sensitivity to model
         parameters as a function of λ.
         λ can be interpreted either as
         the noise constant in eq. 7 or
         underlying the design as the
         only free design parameter in
         the deterministic controller
         defined by eq. 9. Both the gain
         and the filter constant of this
         controller (see Figs. 7,8) are
         functions of λ (see Ref. 28).

The problem that we face is how to
define and delineate our lack of knowledge of G_p. Here we confront a problem also faced by statisticians and by
the chemical reaction engineers in
areas other than control. That is, we
must compensate for the fact that we
use imperfect models, involving uncertainties.

The statistician is familiar with
uncertainties of parameter estimates
and has ways to bound them. However,
the statistician does not have efficient tools to deal with uncertainties
of models themselves.

One way to take care of this problem is to vary the parameters of G_p over a wide range and to demand that the system remains stable. Interestingly, this also often takes care of the fact that we may have not only made an incorrect estimate of the parameters of G_p, but also an incorrect assumption for the functional form of G_p.

Once one overcomes the problems of stability analysis, there is no question that these are superior controllers and that the formulation of Fig. 7 is one of the important recent advances in the design of process controllers.

For control purposes this is, at present, an exciting research area. But my main point here was something else. I wanted to show how the ideas behind modern control theory developed as a result of their formulation as a stochastic estimation problem. While the fundamental idea was generated by a stochastic approach and is an obvious result of it, once we have understood it, we can translate into the language that is familiar to the conventional engineer. Very often researchers are enthusiastic about a mathematical method and object to translating it; they want everybody to learn their language. However, if it can be done there is an advantage to translating the results; it makes them easily understandable to a broad audience. Moreover, sometimes we have an additional side benefit because we gain additional insight from the translation.

Research Challenges for the Future

In the preceding I have given some historic examples of the potential usefulness of probability theory for chemical engineering research and practice. No full review of the subject was intended; I deemed it preferable to illustrate some main ideas. A growing number of young academics is getting familiar with probability, and I would feel presumptuous to tell them what research to do. However, to live up to the title of my talk I want to present some research areas which I think are a real challenge for the profession and could benefit from the introduction of some probabilistic concepts or at least some ideas borrowed from probability. I do not necessarily refer to new

algorithms, but rather to the use of probability to generate new approaches to the central problems of chemical engineering.

One of the present challenges of our profession is to translate the advances of engineering science into better design methods. There is especially a need for this in reaction engineering and control. When we try to do this we are faced with a special problem, namely, that our models are simplified and inaccurate. The more powerful and refined the design procedure, the more important it is to understand the inherent inaccuracies of models and to develop better identification procedures. Thus, for example, the introduction of optimal control into the industry was delayed for years, because it did not take into account sufficiently the problem that the process transfer functions are not well known. Conventional control methods are not seriously affected by model inaccuracies whereas the scheme in Fig. 6b or 7 strongly is.

There are similar problems in reaction engineering processes, in optimization, in linear programming, etc. We use models which, hopefully, incorporate the essential physics and chemistry of the problem but do not describe the system in the way that the Navier Stokes equations describe a fluid dynamic problem. Our models are really refined statistical fitting functions with a sound physical basis[1]. Any advanced design method must be fitted to an identification method. What is needed is a better way to understand the limitations and inaccuracies of model identification methods, especially in complex chemical reactions. If we understand these limiting properties we can also understand the limitations of specific design methods. What we need here is to borrow ideas from the statistician of how to bound our degree of uncertainty in a quantitative way. Once we know how to do this we can take the second step, namely, to match each design method with a proper identification method.

Even if we do not always translate this into quantitative methods of defining model errors, they can educate the engineer to think continuously about model errors and uncertainties and their potential impact on design. This, in itself, can be an important contribution.

REFERENCES

1. Shinnar, R.   Chemical Reactor Modelling, Chemical Reaction Reviews, ACS Symposium Series, No. 72, p. 1 (1978).

2. Deming, W.E.   "Some Theories of Sampling," Wiley, N.Y. (1950).

3. Natrella, M.G.   Experimental Statistics National Bureau of Standards Handbook 91 (1966).

4. Montgomery, D.C.   Design and Analysis of Experiments, John Wiley (1976).

5. Hald, A.   Statistical Theory with Engineering Applications, John Wiley (1952).

6. Brian, P.L.T., and Dolp, L.C. IEC Fundamentals 21, p. 101 (1982).

7. Krambeck, F., Katz, S. and Shinnar, R.   IEC Fundamentals 8, 432(1969).

8. Seinfeld, H.J. and Lapidus, L. Mathematical Methods in Chemical Engineering, Vol. 3, Process Modelling Estimation and Identification, Prentice Hall (1974).

9. Herz, David.   "New power for management systems and management science," McGraw-Hill (1969).

10. Oppenheim, I., Shuler, K.E. and Weiss, G.H.   "The Master Equation in Chemical Physics," MIT Press (1977).

11. Katz, S. and Shinnar, R.   "Probability Methods in Particulate Systems," IEC vol. 61, 60-73 (1969).

12. Feller, William.   "Elements of Theory of Markov Processes and their Applications," McGraw-Hill, N.Y. (1960).

13. Saidel, G.M., Katz, S. and Shinnar, R.   Advances in Chem. Ser. 91, A.C.S., Washington, D.C. 145-157 (1969).

14. Shinnar, R. and Katz, S.   Chemical Reaction Engr., Amer. Chem. Soc. (1972).

15. Hermans, J.J.   Polymer Solution Properties, Dowden, Hutchinson and Ross, publ., Stroudsburg, Pa. (1977).

16. Shah, B., Ramakrishna, D., and Borwanker, J.D.   AIChE Journal 23, 897, 1977.

17. Bayewitz, M., Yerushalmi, J., Katz, S. and Shinnar, R. J. of Atmospheric Sciences 31, No. 6, (1974).

18. Weiss, G.H. and Rubin, R.J. Advances in Chemical Physics 52, 363(1983).

19. Goldsmith, R.P., Amundson, N.R. Chem. Eng. Sci. 195, 559(1965).

20. Hulburt, H.M. and Katz, S.   Chem. Eng. Sci. 19, 555(1964).

21. Smith, W.V. and Ewart, R.H.   J. Chem. Phys. 16, 592(1948).

22. Danckwerts, P.V.   Chem. Eng. Sci. 2, 1(1953).

23. Shinnar, R. and Naor, P.   IEC Fundamentals 2, 278(1963).

24. Nauman, E.B. and Buffham, B.A. Mixing in Continuous Flows, Wiley Interscience (1983).

25. Shinnar, R., Glasser, D. and Katz, S.   Chem. Eng. Sci., Vol. 28, 617-621(1973).

26. Shinnar, R.   "Tracer Experiments in Chemical Reactor Design." Proceedings of first Conference on Physicochemical Hydrodynamics, Advance publishing to London, 1977.

27.   Wiener, N. The Extrapolation
            Interpretation and Smoothing
            of Stationary Time Series,
            Wiley, N.Y. 1949.

28.   Palmor, Z. and Shinnar, R. IEC
            Proc. Des. & Dev. 16, p. 1,
            1978.

29.   Palmor, Z. and Shinnar, R. AIChE
            Journal 27, 793 (1981).

30.   Brosilow, C.B. JACC Proceedings,
            Denver (1979).

31.   Garcia, C.E. and Morari, M. IEC
            Process Design 21, 308(1982).

32.   Kalman, R.E. Trans ASME Basic
            Eng. 82, 35(1960).

33.   Box, G.E.P. and Jenkins, G.M.
            Time Series Analysis, Fore-
            casting and Control, Holden
            Day, San Francisco (1976).

REACTION, DIFFUSION, AND ELEMENTARY FUNCTIONAL ANALYSIS

Neal R. Amundson ■ University of Houston, Houston, TX 77004

The solution of complicated reaction and diffusion problems if attempted by standard techniques soon leads to conceptual difficulties of an almost insurmountable kind. This paper discusses how the formalism for the solution of simple problems can be generalized using ideas of elementary linear functional analysis to solve problems of great complexity. Linear problems in multicomponent reaction and diffusion using Wei-prater kinetics as well as some others for spheres in fixed beds both for the transient and steady state are solved. While the formalism is precise the details problably require computer symbol manipulation to reach a numerical solution.

SCOPE

Multicomponent reaction and diffusion problems either for single particles or for particles in a packed bed or other reactor geometry lead to coupled systems of ordinary or partial differential equations. The method of attack on such problems at first glance is not obvious. However, the use of linear operator theory involving the definition of particular function spaces endowed with the appropriate inner products for which the linear operator is self adjoint frequently brings these problems within reach and is limited to a large extent only by the inability to discover the correct topological structure of the function space. This is not to say that all problems can be so treated since there are problems which are inherently non-self adjoint. For example, if the eigenvalues are complex it is pointless to look for a self adjoint formalism. The idea in this paper is to use systems of first order reactions of the Wei-Prater type which reaction matrix has an inner product on a finite dimensional vector space which renders that matrix a symmetric operator. Some generalizations of this inner product are then used in multicomponent diffusion problems for a single sphere, for a sphere having shells of different catalytic activity, for the steady state in a fixed isothermal bed, and for the transient isothermal fixed bed. These problems all lead to self adjoint eigenvalue problems with real eigenvalues. To be rigorous the Green's functions

University of Houston, Houston, Texas.

for the appropriate differential operator should be generated in order to show the self adjointness of the inverse systems. That is not done here. The computations here are all formal and to develop numerical solutions some help with symbol manipulation should be available since the calculation of eigenvalues and eigenfunctions while conceptually simple is not algebraically or numerically simple.

CONCLUSIONS AND SIGNIFICANCE

The literature is saturated with papers on reaction and diffusion in single particles and in whole reactors, all inspired years ago by the classic work of Ernest Thiele. More and more complicated systems have been solved through the years. The present paper adds to the literature by solving the multicomponent diffusion problem for Knudsen diffusion with Wei-Prater kinetics for some transient problems useful in systems in which temperature effects, such as in some isomerizations, may not be important. Systems with an arbitrary number of chemical species in a packed bed with intraparticle diffusion and axial dispersion are presented. The spherical packing in the bed may consist of an arbitrary number of concentric spherical shells each with its own diffusion and reaction matrix. The methods used are those which have been used for transport in composite media as well as coupled reactor systems in other contexts. Detailed numerical studies are not presented since some form of symbol manipulation is essential for their realization.

INTRODUCTION

The purpose of this paper is to exploit a formalism for some reaction and diffusion problems of similar kind and of linear type which enables one to obtain solutions at least formally by using methods analogous to those used for much simpler problems. These schemes have been exploited in the literature by Ramkrishna and Amundson for vibration problems (1974a) for problems involving transport in composite media (1974), for multicomponent rectification (1973) and for some relatively simple reactor engineering problems (1974b). These problems are linear and therefore of not as great applicability as one might hope but it is true that most non-linear problems are reduced to linear problems and then solved by an iterative procedure. The great joy of linear problems is that analytical solutions can often be brought into reach and the solution structure made explicit. No claim is made here that earthly engineering problems are being solved. Rather, our aim is to make explicit a technique which we hope others might apply to problems of a more pragmatic kind.

The general procedure is best illustrated by considering a simple model problem of reaction and diffusion in a porous sphere. The describing equations are

$$\frac{\varepsilon D}{r^2} \frac{\partial}{\partial r}(r^2 \frac{\partial c}{\partial r}) - kc = \varepsilon \frac{\partial c}{\partial \theta} \quad , \quad 0 < r < a \quad , \quad \theta > 0$$

$$c = 0 \quad , \quad r = a \quad , \quad \theta > 0 \; ; \; c = \tilde{f}(r) \quad , \quad \theta = 0 \quad , \quad 0 < r < a \; .$$

The complication of setting $c = g(\theta)$ at $r = a$ will be avoided at this point although it causes no difficulty. In dimensionless form we have

$$\frac{1}{\rho^2} \frac{\partial}{\partial \rho}(\rho^2 \frac{\partial y}{\partial r}) - \beta y = \frac{\partial y}{\partial \tau} \quad , \quad 0 < \rho < 1 \quad , \quad \tau > 0 \quad (1)$$

$$y = 0 \quad , \quad \rho = 1 \quad , \quad \tau > 0 \; ; \; y = F(\rho) \quad , \quad \tau = 0 \quad , \quad 0 < \rho < 1. \quad (2,3)$$

We consider the eigenvalue problem

$$-\frac{1}{\rho^2} \frac{d}{d\rho}(\rho^2 \frac{dw}{d\rho}) + \beta w = \lambda w$$

$$w = 0 \quad , \quad \rho = 1 \; ; \; w \text{ finite at } \rho = 0.$$

This clearly generates a complete set of orthonormal eigenfunctions with a discrete set of eigenvalues as

$$\{\lambda_n = n^2 \pi^2 + \beta, w_n = \sqrt{2}\frac{\sin n\pi\rho}{\rho} \quad , \quad 0 < \rho < 1\}.$$

If we use the usual inner product, defined on the appropriate Hilbert space,

$$(u,v) = {}_0\!\int^1 \rho^2 u(\rho)v(\rho)d\rho$$

then we know that any function $f(\rho)$ in $L_2(0,1)$ has a representation

$$f(\rho) = {}^{\infty}_1\!\Sigma(f,w_n)w_n(\rho) \quad (4)$$

which converges in the mean, and where

$$(f,w_n) = {}_0\!\int^1 \rho^2 f(\rho)w_n(\rho)d\rho \; . \quad (5)$$

Equations (4) and (5) may be treated as Fourier transform pairs. We will call (f,w_n) the Finite Fourier Transform (FFT) of $f(\rho)$ with respect to the set $\{w_n(\rho)\}$ on $(0,1)$. So

$$F[f(\rho)] = (f,w_n) = {}_0\!\int^1 \rho^2 f(\rho)w_n(\rho)d\rho$$

and it follows that the inverse of F is

$$F^{-1}[(f,w_n)] = f(\rho) = {}^{\infty}_1\!\Sigma(f,w_n)w_n(\rho).$$

We can now apply this transform to the set of Equations (1), (2), and (3) as

$$F[\frac{1}{\rho^2} \frac{\partial}{\partial \rho}(\rho^2 \frac{\partial y}{\partial \rho}) - \beta y] = F[\frac{\partial y}{\partial \tau}]$$

to obtain

$$(-n^2\pi^2 - \beta)y_n = \frac{dy_n}{d\tau}$$

with the result

$$y(\rho,\tau) = {}^{\infty}_1\!\Sigma(F,w_n)e^{-(n^2\pi^2+\beta)\tau}w_n(\rho).$$

Now the reason this procedure is so clean is that the eigenvalue problem is a self adjoint one (although not regular) producing real eigenvalues, a complete set of orthonormal eigenfunctions, and an expansion theorem for functions of a broad class. This results from the fact that while the differential operator is unbounded it does have a symmetric Green's function so that the inverse of the differential system is bounded and is indeed self adjoint. The reader is referred to Coddington and Levinson (1955), Cole (1968) or to Ramkrishna and Amundson (1985). For this procedure to be valid one must show the existence of the Green's functions. We will not do this in the sequel but it can be done, even if very tediously.

In the event that $c = g(\vartheta)$, $r = a$, or $y = G(\tau)$, $\rho = 1$, then the solution may be easily shown to be

$$y(\rho,\tau) = {}^{\infty}_1\!\Sigma(F,w_n)e^{-(n^2\pi^2+\beta)\tau}w_n(\rho)$$

$$- {}^{\infty}_1\!\Sigma w_n(\rho)w_n'(1){}_0\!\int^\tau e^{-(\tau-\mu)(n^2\pi^2+\beta)}G(\mu)d\mu.$$

More complicated boundary conditions can be treated as well.

MULTICOMPONENT REACTION AND DIFFUSION IN A SPHERE

Our first problem is one in which we have the general reaction system

$$A_i \underset{i k_{ij}}{\overset{k_{ji}}{\rightleftarrows}} A_j \quad , \quad i,j = 1 \text{ to } n \quad ,$$

exploited by Wei and Prater (1962). The reaction matrix is given by

$$\bar{\bar{K}}=[k_{ij}] \quad , \quad k_{jj}=-\sum_{i=1,j}^{n}k_{ij} \quad , \quad i,j=1 \text{ to } n \quad . \qquad (6)$$

Considered as an operator on R^n, the matrix $\bar{\bar{K}}$ is symmetric with respect to the inner product (as shown by Ramkrishna and Amundson (1973))

$$<\bar{u},\bar{v}>_1=\sum_{j=1}^{n}\frac{u_i v_i}{a_i} \quad ; \quad <\bar{\bar{K}}\bar{u},\bar{v}>_1=<\bar{u},\bar{\bar{K}}\bar{v}>_1 \qquad (7)$$

where the a_i are the equilibrium concentrations obtained from $k_{ji}a_i=k_{ij}a_j$ necessitated by the principle of detailed balancing, and these depend only upon the temperature.

We consider now a generalization of our model problem

$$\frac{\varepsilon\bar{\bar{D}}}{r^2}\frac{\partial}{\partial r}(r\frac{\partial\bar{c}}{\partial r})+\gamma\bar{\bar{K}}\bar{c}=\varepsilon\frac{\partial\bar{c}}{\partial\theta} \quad , \quad 0<r<a \quad , \quad \theta>0 \qquad (8)$$

$$\bar{c}=0 \quad , \quad r=a, \quad \theta>0 \quad ; \quad \bar{c}=\bar{f}(r) \quad , \quad \theta=0 \quad , \quad 0<r<a$$

where $\bar{\bar{D}}$ is a diagonal matrix and $\bar{\bar{K}}$ is the Wei-Prater matrix, Equation (6), of reaction velocity constants. As above we consider the operator

$$L\bar{u}=-\frac{\varepsilon\bar{\bar{D}}}{r^2}\frac{d}{dr}(r^2\frac{d\bar{u}}{dr})-\gamma\bar{\bar{K}}\bar{u} \qquad (9)$$

which operates on n component vectors each component being a concentration of an A_i in the reaction scheme and each of which is a function of the space variable r. Defining an inner product on this space

$$<\bar{u},\bar{v}>_2=\sum_{1}^{n}\int_0^a r^2\frac{u_i v_i}{a_i}dr$$

gives us a Hilbert space on which it may be shown that the operator Equation (9) is self adjoint and positive definite (Ramkrishna and Amundson (1985)). We now consider the eigenvalue problem,

$$-\frac{\varepsilon\bar{\bar{D}}}{r^2}\frac{d}{dr}(r^2\frac{d\bar{w}}{dr})-\gamma\bar{\bar{K}}\bar{w}=\varepsilon\lambda\bar{w} \quad , \quad 0<r<a$$

$$\bar{w}=0 \quad , \quad r=a \quad ; \quad \bar{w} \text{ bounded}$$

where from the positive definitness, all of the eigenvalues $\{\lambda_n\}$ must be positive, and write it in the form

$$\frac{1}{r^2}\frac{d}{dr}(r^2\frac{d\bar{w}}{dr})+(\frac{\gamma}{\varepsilon}\bar{\bar{D}}^{-1}\bar{\bar{K}}+\bar{\bar{D}}^{-1}\lambda)\bar{w}=0$$

We consider the matrix

$$\frac{\gamma}{\varepsilon}\bar{\bar{D}}^{-1}\bar{\bar{K}}+\bar{\bar{D}}^{-1}\lambda=[\frac{\gamma}{\varepsilon}\frac{k_{ij}}{D_i}+\frac{\lambda}{D_i}\delta_{ij}] \quad .$$

This matrix is symmetric with respect to the inner product

$$<\bar{u},\bar{v}>_3=\sum_{1}^{n}\frac{D_i u_i v_i}{a_i}$$

so that its eigenvalues are real but the oper-

ator with that inner product is not positive definite. Let

$$\bar{\bar{\Phi}}^2=\frac{\gamma}{\varepsilon}\bar{\bar{D}}^{-1}\bar{\bar{K}}+\bar{\bar{D}}^{-1}\lambda$$

so that

$$\frac{1}{r^2}\frac{d}{dr}(r^2\frac{d\bar{w}}{dr})+\bar{\bar{\Phi}}^2\bar{w}=\bar{0}$$

and the solution satisfying the boundedness condition is

$$\bar{w}=\bar{\bar{\Phi}}^{-1}\frac{\sin\bar{\bar{\Phi}}r}{r}\bar{k} \quad .$$

Note that in this solution the square root of $\bar{\bar{\Phi}}^2$ does not really appear since

$$\bar{w}=[\bar{\bar{I}}-\frac{\bar{\bar{\Phi}}^2 r^2}{3!}+\frac{\bar{\bar{\Phi}}^4 r^4}{5!}-\frac{\bar{\bar{\Phi}}^6 r^6}{7!}+\ldots]\bar{k} \quad .$$

At r=a we must have

$$\bar{\bar{\Phi}}^{-1}\sin\bar{\bar{\Phi}}a\bar{k}=\bar{0} \quad .$$

Using the spectral theorem we can write

$$\bar{\bar{\Phi}}^{-1}\sin\bar{\bar{\Phi}}a\bar{k}=\sum_{j=1}^{n}\frac{\sin\sqrt{\eta_j}a}{\sqrt{\eta_j}}\frac{\text{adj}(\eta_j\bar{\bar{I}}-\bar{\bar{\Phi}}^2)}{\prod_{i=1,j}^{n}(\eta_j-\eta_i)}\bar{k}$$

where η_j is an eigenvalue of $\bar{\bar{\Phi}}^2$. For a non-trivial solution

$$\det[\bar{\bar{\Phi}}^{-1}\sin\bar{\bar{\Phi}}a]=0$$

which is an equation in λ. Now we know the λ's must be positive so to compute them we assume a value for λ and calculate the eigenvalues of $\bar{\bar{\Phi}}^2$ corresponding. Whether this was the right value for λ or not is determined by the above spectral resolution. An iterative procedure of some kind then seems essential. We can then compute the appropriate $\bar{k}$, call it $\bar{k}_n$ to obtain

$$\bar{w}_n=a\bar{\bar{\Phi}}^{-1}\frac{\sin\bar{\bar{\Phi}}r}{r}\bar{k}_n \quad .$$

We assume now from the general theory that the set $\{\bar{w}_n\}$ is complete in the appropriate n-vector function space on $l_2[0,a]$ and that we have the orthogonality property

$$<\bar{w}_n,\bar{w}_m>_2=\int_0^a\rho^2\sum_{1}^{n}\frac{w_{ni}w_{mi}}{a_i}d\rho=\delta_{nm} \quad .$$

We note that $\bar{w}_n$ can be written in the form

$$\bar{w}_n=\frac{\sin\sqrt{\eta_{jn}}r}{\sqrt{\eta_{jn}}}\bar{z}_{jn}$$

so that the ith component of $\bar{w}_n$ is

$$w_{ni}=\sum_{j=1}^{n}\frac{\sin\sqrt{\eta_{jn}}r}{\sqrt{\eta_{jn}}r}z_{jni} \quad .$$

In the event that some of the η_{jn} are negative, the hyperbolic sine will appear. We are now in a position to define a FFT as we did with Equations (4) and (5)

$$F[\bar{u}(\rho,t)] = \int_0^a \sum_1^w \frac{u_{ni} u_i}{a_i} \rho^2 d\rho = u_n(t).$$

Application to the original differential Equation (8) gives

$$F[L\bar{c}] = -\varepsilon F[\frac{\partial \bar{c}}{\partial \theta}]$$

$$\varepsilon \frac{dc_n}{d\theta} = -\lambda_n c_n$$

or

$$\bar{c}(r,\theta) = \sum_1^\infty F[\bar{f}(r)] e^{-\frac{\lambda_n \theta}{\varepsilon}} \bar{w}_n .$$

Now it is clear from the above analysis that while the formalism is straightforward the execution of the details is tedious and computer symbol manipulation probably essential. It is simpler to proceed as follows with the same problem but with boundary conditions.

$$-\varepsilon \bar{\bar{D}} \frac{\partial \bar{c}}{\partial r})_{r=a} = \bar{\bar{k}}[\bar{c} - \bar{c}_0(\theta)] , \quad r=a$$

and initial condition $\bar{c} = \bar{f}(r)$, $\theta = 0$. We assume that $\bar{\bar{D}}$ and $\bar{\bar{k}}$ are diagonal and that $k_j/D_j = S$, a constant independent of j. This simplifies things but may not be essential. We are still considering Equation (8). If we consider the scalar FFT derived from the eigenvalue problem

$$-\frac{1}{r^2} \frac{d}{dr}(r^2 \frac{dz}{dr}) = \lambda z , \quad 0 < r < a$$

$$-\frac{dz}{dr} = Sz , \quad S > 0 , \quad r=a$$

we obtain a set of eigenvalues λ_n (not the same as those above) and an orthonormal set of eigenfunctions $\{z_n(r)\}$ which give us a transform pair which when applied to the system Equation (8) gives

$$\varepsilon \frac{d\bar{c}_n}{d\theta} = (-\varepsilon \bar{\bar{D}}\lambda_n + \gamma \bar{\bar{K}})\bar{c}_n + a^2 \bar{\bar{k}}\bar{c}_0(\theta) z_n(a)$$

$$\bar{c}_n = \bar{f}_n = F[\bar{f}(r)] .$$

We can obtain such a form since each of the chemical species is involved in the equation in the same way. Had k_j/D_j been a function of j the procedure would not have been so straightforward. This is a set of first order linear equations for which the matrix $-\varepsilon \bar{\bar{D}}\lambda_n + \gamma \bar{\bar{K}}$ is symmetric with respect to the inner product

$$\langle \bar{u}, \bar{v} \rangle_1 = \sum_1^n \frac{u_i v_i}{a_i}$$

and the eigenvalues and eigenvectors defined

by

$$(-\varepsilon \bar{\bar{D}}\lambda_n + \gamma \bar{\bar{K}})\bar{\omega}_{nj} = \eta_{nj}\bar{\omega}_{nj}$$

are negative and orthonormal, respectively, so that a little fourier transform ($\oint$) and eigenvector expansion can be defined. Let

$$\oint(\bar{c}_n) = c_{nj}$$

and

$$\bar{c}_n = \oint^{-1}(c_{nj}) = \sum_{j=1}^n \oint(\bar{c}_n)\bar{\omega}_{nj} .$$

Direct application gives

$$\frac{dc_{nj}}{d\theta} = \eta_{nj}c_{nj} + \langle \bar{\bar{k}}\bar{c}_0(r)z_n(a), \bar{\omega}_{nj}\rangle_1$$

$$c_{nj} = f_{nj} = \langle \bar{f}_n, \bar{\omega}_{nj}\rangle_1 , \quad \theta=0$$

and the solution is then

$$c_{nj} = f_{nj} e^{\eta_{nj}\theta} + a^2 \int_0^\theta e^{\eta_{nj}(\theta-\eta)} \langle \bar{\bar{k}}\bar{c}_0(\eta), \bar{\omega}_{nj}\rangle_1 z_n(a)d\eta$$

so that

$$\bar{c}(r,\theta) = \sum_1^\infty \sum_1^n f_{nj} e^{\eta_{nj}\theta} z_n(r)\bar{\omega}_{nj}$$

$$+ a^2 \sum_1^\infty \sum_1^n z_n(r)\bar{\omega}_{nj} z_n(a) \int_0^\theta e^{\eta_{nj}(\theta-\eta)} \langle \bar{\bar{k}}\bar{c}_0(\eta), \bar{\omega}_{nj}\rangle_1 d\eta .$$

This solution will certainly be much easier to apply. The steady state versions of this problem appear in Aris (1975) and Ramkrishna and Amundson (1985).

COMPOSITE MEDIA

The problem above can be solved in principle for a composite sphere with shells of different catalytic activity and porosity. Suppose in each shell, $r_{j-1} < r < r_j$, $j=1$ to s, $r_s = a$, that the catalytic activity is described by a Wei-Prater matrix $\bar{\bar{K}}^j$ and a diagonal diffusion matrix $\bar{\bar{D}}^j$ and porosity ε_j. At the shell interfaces compatibility conditions are required on the concentrations and fluxes. The model is

$$\varepsilon_j \frac{\bar{\bar{D}}^j}{r^2} \frac{\partial}{\partial r}(r^2 \frac{\partial \bar{c}}{\partial r}) + \gamma_j \bar{\bar{K}}^j \bar{c} = \varepsilon_j \frac{\partial \bar{c}}{\partial \theta} , \quad r_{j-1} < r < r_j, \quad j=1 \text{ to } s$$

$$\bar{c}(r_j-,\theta) = \bar{c}(r_j+,\theta) , \quad j=1 \text{ to } s-1$$

$$\varepsilon_{j-1} \bar{\bar{D}}^{j-1} \frac{\partial \bar{c}}{\partial r})_{r_j-} = \varepsilon_j \bar{\bar{D}}^j \frac{\partial \bar{c}}{\partial r})_{r_j+} , \quad j=1 \text{ to } s-1$$

$$-\varepsilon_s \bar{\bar{D}}^s \frac{\partial \bar{c}}{\partial r})_{a-} = \bar{\bar{k}}(\bar{c} - \bar{c}^0(\theta)) , \quad r=r_s=a$$

$\bar{c}$ bounded at $r=0$; $\bar{c}=\bar{f}(r)$, $\theta=0$.

We point out that $\bar{c}$ is a vector of n components, each component representing a species A_i so that these vectors really have components c_{ij}, $i=1$ to $n, j=1$ to s. We should write $\bar{c}_j$ for $\bar{c}$ then the compatibility condition would be $\bar{c}_{j-1}(r_j-,\theta)=\bar{c}_j(r_j+,\theta)$. We now consider the operator

$$L\bar{u}=-\frac{\bar{\bar{D}}^j}{r^2}\frac{d}{dr}(r^2\frac{d\bar{u}}{dr})-\frac{\gamma_j}{\varepsilon_j}\bar{\bar{K}}^j\bar{u} \ , \ r_{j-1}<r<r_j \ , \ j=1 \text{ to } s$$

$$\bar{u})_{r_j-}=\bar{u})_{r_j+}$$

$$\varepsilon_{j-1}\bar{\bar{D}}^{j-1}\frac{d\bar{u}}{dr})_{r_j-}=\varepsilon_j\bar{\bar{D}}^j\frac{d\bar{u}}{dr})_{r_j+}$$

$$-\varepsilon_s\bar{\bar{D}}^s\frac{d\bar{u}}{dr})_{a-}=\bar{\bar{k}}\bar{u} \ , \ r=a \ .$$

While this appears formidable it can be shown to be self adjoint and positive definite with respect to the inner product

$$<\bar{u},\bar{v}>_4=\sum_{j=1}^{s}\varepsilon_j\int_{r_{j-1}}^{r_j}r^2\sum_{i=1}^{n}\frac{u_i^jv_i^j}{a_i}dr$$

in a suitable Hilbert space. While we write $\bar{u}$ and $\bar{v}$ in the inner product it is clear that the inner product is over s vectors $\bar{u}$ and s vectors $\bar{v}$ in the spherical shells. The a_i are the same in each shell because of the isothemicity. The eigenvalue problem $L\bar{u}=\lambda\bar{u}$ then has real positive eigenvalues and an orthonormal set of eigenvectors and the general theory then gives us an expansion theorem for function vectors, $\bar{f}(r)$, vectors with n components in each shell.

$$\bar{f}(r)=\sum_1^{\infty}\alpha_m\bar{u}_m(r)$$

$$\alpha_m=<\bar{f}(r),\bar{u}_m>$$

but where we note that $\bar{f}(r)$ and $\bar{u}_m$ have different definitions in each shell. As before we can now apply the FFT to the original system. Thus

$$F[\bar{c}]=<\bar{c},\bar{u}_m>_4=c_m(\theta)$$

and application to the system gives

$$-\lambda_m c_m+a^2\sum_1^{n}\frac{k_ic_i(\theta)u_{im}(a)}{a_i}=\frac{dc_m}{d\theta}$$

where $u_{im}(a)$ is the ith component of $\bar{u}_m$ evaluated at $r=a$. With this equation we associate $\bar{c}=\bar{f}(r)$ when $\theta=0$ so that the final solution is

$$\bar{c}(r,\theta)=\sum_{n=1}^{\infty}e^{-\lambda_m\theta}<\bar{f}(r),\bar{u}_m>_4\bar{u}_m(r)$$

$$+a^2\sum_{n=1}^{\infty}\bar{u}_m(r)\int_0^{\theta}e^{-\lambda_m(\theta-\tau)}\sum_{i=1}^{n}\frac{k_ic_i(\tau)u_{im}(a)}{a_i}d\tau \ .$$

To use such a formula would not be a trivial task since the computation of the eigenvalues and eigenfunctions would be a non-trivial exercise. Some form of computer symbol manipulation would be essential.

FIXED BED REACTOR--TRANSIENT

We now consider a fixed bed reactor of spherical catalyst particles in which concentric spherical shells within the particles contain catalyst of different activities. Suppose there are s such shells with radii $0<r_1<r_2<r_3<...<r_s=a$. Suppose also the diffusion matrix which is diagonal is different in each shell. Call it $\bar{\bar{D}}_j$, $j=1,2,...,s$ and that the porosity ε_j is also different for each j. Again we suppose we have the Wei-Prater kinetics in each shell so that the kinetic matrix is $\bar{\bar{K}}^j$. At the boundaries of the shells there will be compatibility conditions of the usual kind for transport systems. At the particle surface there may be a mass transfer resistance between that surface and the intersticial fluid. The usual equations then are

$$\frac{\varepsilon_j\bar{\bar{D}}^j}{r^2}\frac{\partial}{\partial r}(r^2\frac{\partial\bar{c}_j}{\partial r})+\alpha_j\bar{\bar{K}}^j\bar{c}_j=\varepsilon_j\frac{\partial\bar{c}_j}{\partial\theta} \ ; \quad \begin{matrix} r_{j-1}<r<r_j \\ j=1,2,...,s-1 \end{matrix}$$

$$\bar{c}_{j-1}(r_j-,\theta)=\bar{c}_j(r_j+,\theta)$$
$$j=1,2,...,s-1$$

$$\varepsilon_{j-1}\bar{\bar{D}}^{j-1}\frac{\partial\bar{c}_j}{\partial r})_{r_j-}=\varepsilon_j\bar{\bar{D}}^j\frac{\partial\bar{c}_j}{\partial r})_{r_j+}$$

$$-\varepsilon_s\bar{\bar{D}}^s\frac{\partial\bar{c}_s}{\partial r})_{r=a}=\bar{\bar{k}}(\bar{c}_s-\bar{C}) \ , \ r=a$$

$$\bar{c}_j=\bar{f}_j(r) \ , \ \theta=0 \ , \ j=1,2,...,s$$

where $\bar{c}$ is the concentration in the interstitial fluid. We assume that in the intersticial fluid all species behave in the same way; that is they are all transported with the same speed so that the intersticial velocity u and axial dispersion D do not depend upon i, the species number. For the intersticial fluid then we have

$$\gamma D\frac{\partial^2\bar{C}}{\partial x^2}-u\gamma\frac{\partial\bar{C}}{\partial x}-4\pi a^2 ND\bar{\bar{\varepsilon}}^s\frac{\partial\bar{c}_s}{\partial r})_{r=a}=\gamma\frac{\partial\bar{C}}{\partial\theta}$$

where N is the number of particles per unit of volume. This equation can be written as

$$D\frac{\partial^2 \bar{C}}{\partial x^2} - u\frac{\partial \bar{C}}{\partial x} - \beta\bar{\bar{D}}^s\frac{\partial \bar{c}_s}{\partial r})_{r=a} = \frac{\partial \bar{C}}{\partial \theta}$$

where

$$\beta = \frac{4\pi a^2 N}{\gamma} = \frac{3(1-\gamma)}{\gamma a} .$$

With this equation we need the two usual boundary conditions

$$-D\frac{\partial \bar{C}}{\partial x} + u\bar{C} = u\bar{C}_{in}(\theta) , \quad x=0 , \quad \theta>0$$

$$\frac{\partial \bar{C}}{\partial x} = 0 , \quad x=\ell , \quad \theta>0 .$$

We will not reduce this system to dimensionless form since it helps but little in this problem and obscures some of the details.

Our solution strategy is to apply a FFT to remove the dependence on the x variable and then to try to redefine the linear operator on r and an associated inner product on a suitable Hilbert space that will render that operator self adjoint. The necessity of showing the existence of a symmetric Green's function will be avoided but it is assumed to exist.

It is convenient, although not essential, to make the substitutions

$$\bar{c}_j = \bar{v}_j e^{\frac{ux}{2D}} ; \quad \bar{C} = \bar{V}e^{\frac{ux}{2D}}$$

in the above equations. Since the variable x in the particle equations is only implicit, they retain their form while the equations containing $\bar{C}$ change theirs. We have

$$\frac{\varepsilon_j\bar{\bar{D}}^j}{r^2}\frac{\partial}{\partial r}(r^2\frac{\partial \bar{v}_j}{\partial r}) + \alpha_j\bar{\bar{K}}^j\bar{v}_j = \varepsilon_j\frac{\partial \bar{c}_j}{\partial \theta}$$

$$\bar{v}_{j-1}(r_j-,\theta)\bar{v}_j(r_j+,\theta)$$

$$\varepsilon_{j-1}\bar{D}^{j-1}\frac{\partial \bar{v}_{j-1}}{\partial r})_{r_j-} = \varepsilon_j\bar{D}^j\frac{\partial v_j}{\partial r})_{r_j+} \qquad j=1,2,3,\ldots,s-1$$

$$-\varepsilon_s\bar{\bar{D}}^s\frac{\partial \bar{v}_s}{\partial r})_{r=a} = \bar{\bar{k}}(\bar{c}_s-\bar{C}) , \quad r=a$$

$$\bar{v}_j = \bar{f}_j(r)e^{-\frac{ux}{2D}} , \quad \theta=0 , \quad j=1,2,\ldots,s$$

and for the intersticial fluid

$$D\frac{\partial^2 \bar{V}}{\partial x^2} - \frac{u^2}{4D}\bar{V} - \beta\bar{\bar{D}}^s\frac{\partial \bar{v}_s}{\partial r})_{r=a} = \frac{\partial \bar{V}}{\partial \theta}$$

$$-D\frac{\partial \bar{V}}{\partial x} + \frac{u}{2}\bar{V} = u\bar{C}_{in}(\theta) , \quad x=0$$

$$D\frac{\partial \bar{V}}{\partial x} + \frac{u}{2}\bar{V} = 0 , \quad x=\ell$$

$$\bar{C} = \bar{F}(x)e^{\frac{ux}{2D}} , \quad \theta=0 , \quad \theta<x<\ell .$$

In this form the differential operator on $\bar{V}$ with respect to x is formally self adjoint and with these boundary conditions forms a self adjoint system. It then follows that the scalar eigenvalue problem

$$D\frac{d^2 w}{dx^2} = -\lambda w$$

$$D\frac{dw}{dx} - \frac{u}{2}w = 0 , \quad x=0$$

$$D\frac{dw}{dx} + \frac{u}{2}w = 0 , \quad x=\ell$$

is a Sturm-Liouville system and so the eigenvalues are real and positive and there is a complete set of orthonormal eigenfunctions w_j with the associated expansion theorem for functions belonging to $L_2(0,\ell)$. Thus we can define a FFT on x. Let

$$F(\bar{v}_j) = \int_0^\ell \bar{v}_j w_k dx = \bar{v}_{jk}$$

and

$$\bar{v}_j = \sum_{k=1}^\infty \bar{v}_{jk}w_k$$

Since this is a scalar transform, its application to a vector gives a vector. Each component species is operated on with the same transformation. Also

$$F(\bar{V}) = \int_0^\ell \bar{V}w_1 dx = \bar{V}_k .$$

The equations then become

$$\frac{\varepsilon_j\bar{\bar{D}}^j}{r^2}\frac{\partial}{\partial r}(r^2\frac{\partial \bar{v}_{jk}}{\partial r}) + \alpha_j\bar{\bar{K}}^j\bar{v}_{jk} = \varepsilon_j\frac{\partial \bar{v}_{jk}}{\partial \theta}$$

$$\bar{v}_{j-1,k}(r_j-,\theta) = v_{jk}(r_j+,\theta) \qquad j=1,2,\ldots,s-1$$

$$\varepsilon_{j-1}\bar{D}^{j-1}\frac{\partial \bar{v}_{j-1,k}}{\partial r})_{r_j-} = \varepsilon_j\bar{D}^j\frac{\partial \bar{v}_{jk}}{\partial r})_{r_j+}$$

$$-\varepsilon_s\bar{\bar{D}}^s\frac{\partial \bar{v}_{sk}}{\partial r})_{r=a} = \bar{\bar{k}}(\bar{v}_{sk}-\bar{V}_k) , \quad r=a \qquad (10)$$

$$\bar{v}_{jk} = \bar{f}(r)\delta_k , \quad \theta=0$$

where δ_k is the transform of $\exp(-ux/2D)$ and

$$(-\lambda_k - \frac{u^2}{4D})\bar{V}_k - \beta\bar{\bar{D}}^s\frac{\partial \bar{v}_{sk}}{\partial r})_{r=a} + w_k(0)u\bar{C}_{in}(\theta) = \frac{\partial \bar{V}_k}{\partial \theta} \qquad (11)$$

$$\bar{V}_k = \bar{F}_k = \int_0^\ell \bar{F}(x)\exp(\frac{ux}{2D})w_k dx$$

We observe now that the spherical equation is for transport in composite media but the problem is complicated by the fact that the normal derivative at the surface of the sphere accounting for the flux of species into the intersticial volume occurs also in the equation for $\bar{v}_k$. While this problem has not been solved, similar ones have and so we will proceed to define a new operator and inner product. First we will eliminate the variable $\bar{v}_k$ from the whole system. We consider Equations (10) and (11) and solve the first for $\bar{v}_k$ to obtain

$$\bar{v}_k = \varepsilon_s \bar{\bar{k}}^{-1} \bar{\bar{D}}^s \frac{\bar{v}_{sk}}{r})_{r=a} + \bar{v}_{sk}$$

and differentiate it with respect to

$$\frac{\partial \bar{v}_k}{\partial \theta} = \varepsilon_s \bar{\bar{k}}^{-1} \bar{\bar{D}}^s \frac{\partial}{\partial \theta} (\frac{\partial \bar{v}_{sk}}{\partial r})_{r=a} + \frac{\partial \bar{v}_{sk}}{\partial \theta}$$

and substitute into Equation (11) to obtain finally

$$-\bar{\bar{\beta}}_k^s \frac{\partial \bar{v}_{sk}}{\partial r})_{r=a} - \mu_k' \bar{v}_{sk} = \frac{\partial}{\partial \theta} [\varepsilon_s \bar{\bar{k}}^{-1} \bar{\bar{D}}^s + \bar{v}_{sk}] - w_k(0) u \bar{C}_{in}(\theta)$$

$$\bar{\bar{\beta}}_k^s = (\lambda_k + \frac{u^2}{rD}) \varepsilon_s \bar{\bar{k}}^{-1} \bar{\bar{D}}^s + \beta \bar{\bar{D}}^s = [\beta_{ik}^s]$$

$$\mu_k' = \lambda_k + \frac{u^2}{4D} .$$

A new operator will be defined as

$$L \begin{bmatrix} \bar{v}_{jk} \\ \varepsilon_s \bar{\bar{k}}^{-1} \bar{\bar{D}}^s \frac{\partial \bar{v}_{sk}}{\partial r})_{r=a} + \bar{v}_{sk})_{r=a} \end{bmatrix} \quad j=1,2,3\ldots,s$$

$$= \begin{bmatrix} -\frac{\bar{\bar{D}}^j}{r^2} \frac{\partial}{\partial r}(r^2 \frac{\partial \bar{v}_{jk}}{\partial r}) - \frac{\alpha}{\varepsilon_j} \bar{\bar{k}}^j \bar{v}_{jk} \\ \bar{\bar{\beta}}_k^s \frac{\partial \bar{v}_{sk}}{\partial r})_{r=a} + \mu_k' \bar{v}_{sk})_{r=a} \end{bmatrix}$$

Why one defines the operator in this way is not immediately obvious and its form is a result of trial and error and experience with it and various inner products until one obtains a useful result. There are probably other formulations which may be simpler and better. We note that the r.h.s. of the opera tor is more complicated than it looks at first glance since the first element is really s elements each holding in a region $r_{j-1} < r < r_j$, $j=1$ to s. The second element is evaluated at $r=a$.

After some manipulation and experimentation it will be found that the appropriate inner product on the appropriate Hilbert space is of the form

$$(\tilde{u}, \tilde{v}) = \sum_{j=1}^{s} \varepsilon_j \int_{r_{j-1}}^{r_j} r^2 \sum_{i=1}^{n} \frac{u_{ij} v_{ij}}{a_i} dr + \sum_{i=1}^{n} \gamma_i \frac{u_{is} v_{is}}{a_i}$$

The notation here requires some explanation. The $\tilde{u}$ and $\tilde{v}$ are vectors of species and shell sections since we are attempting to solve the whole sphere problem at once rather than section by section. We need to determine whether there exists a set of positive $[\gamma_i]$ so that the above will be a valid inner product. A further word about the $[a_i]$ is essential. We observe that there is no subscript j as a_{ij}, since although the kinetic matrices $\bar{\bar{K}}^j$ are different, the particle is isothermal and so irrespective of the fact that the rate constants are different, thermodynamics requires that the equilibrium constants and compositions all be the same, being a function of temperature only.

Direct substitution will show that if

$$(L\tilde{u}, \tilde{v}) = (\tilde{u}, L\tilde{v})$$

then

$$(\tilde{u}, \tilde{v}) = \sum_{j=1}^{s} \varepsilon_j \int_{r_{j-1}}^{r_j} r^2 \sum_{1}^{n} \frac{u_{ij} v_{ij}}{q_i} dr + \varepsilon_s a^2 \sum_{1}^{n} \frac{D_i^s}{\beta_{ik}^s} \frac{u_{is} v_{is}}{a_i}$$

where we note that $\bar{\bar{\beta}}_k^s$ is a diagonal matrix whose elements we call β_{ik}^s and $\lambda_k + u^2/4D$ are the elements of a diagonal matrix all of whose elements are the same. We point out that the vectors $\bar{v}_{sk}$ have s and k fixed so that these are species vectors. Thus the global operator is self adjoint, assuming as always that there exists a symmetric Green's function.

We can compute $(L\tilde{u}, \tilde{u})$ and we obtain after some manipulation

$$(L\tilde{u}, \tilde{u}) = -\sum_{j=1}^{s} \varepsilon_j \int_{r_{j-1}}^{r_j} r^2 \sum_{i=1}^{n} \frac{D_i^j}{a_i r^2} \frac{\partial}{\partial r}(r^2 \frac{\partial u_{ijk}}{\partial r}) u_{ijk} dr$$

$$+ \varepsilon_s a^2 \sum_{1}^{n} \frac{D_i^s}{\beta_{ik}^s a_i} [\beta_{ik}^s \frac{\partial u_{isk}}{\partial r})_{r=a} + \mu_k' u_{isk}] u_{isk}$$

$$= \sum_{j=1}^{s} \varepsilon_j \int_{r_{j-1}}^{r_j} r^2 \sum_{i=1}^{n} (\frac{\partial u_{ijk}}{\partial r})^2 \frac{D_i}{a_i} dr + \varepsilon_s a^2 \sum_{i}^{n} \frac{D_i^s}{\beta_{ik}^s a_i} \mu_k' u_{isk}^2 > 0$$

and hence the operator is positive definite and the eigenvalues are all positive.

We can now write the system as

$$L \begin{bmatrix} \bar{v}_{jk} , \; j=1,2,3,\ldots s \\ \varepsilon_s \bar{\bar{k}}^{-1} \bar{\bar{D}}^s \frac{\partial \bar{v}_{sk}}{\partial r})_{r=a} + \bar{v}_{sk})_{r=a} \end{bmatrix}$$

$$= \begin{bmatrix} -\dfrac{\bar{\bar{D}}j}{r^2}\dfrac{\partial}{\partial r}(r^2\dfrac{\partial \bar{v}_{jk}}{\partial r}) - \dfrac{\alpha}{\varepsilon_j}\bar{\bar{K}}j\bar{v}_{jk} \\[12pt] \bar{\bar{\beta}}s_k\dfrac{\partial \bar{v}_{sk}}{\partial r})_{r=a} + \mu'_k\bar{v}_{sk})_{r=a} \end{bmatrix}$$

$$= -\dfrac{\partial}{\partial \theta}\begin{bmatrix} \bar{v}_{jk} \;,\; j=1,2,3,\dots,s \\[10pt] \varepsilon_s\bar{\bar{k}}^{-1}\bar{\bar{D}}s\dfrac{\partial \bar{v}_{sk}}{\partial r})_{r=a} + \bar{v}_{sk})_{r=a} \end{bmatrix} + \begin{bmatrix} 0 \\[10pt] w_k(0)\,u\,\bar{C}_{in}(\theta) \end{bmatrix}$$

We consider now the eigenvalue problem

$$L(\tilde{\bar{\mu}}_j) = L\begin{bmatrix} \bar{\mu}_j \;,\; j=1,2,3,\dots,s \\[10pt] \varepsilon_s\bar{\bar{k}}^{-1}\bar{\bar{D}}s\dfrac{d\bar{\mu}_s}{dr})_{r=a} + \mu_s)_{r=a} \end{bmatrix}$$

$$= \eta\begin{bmatrix} \bar{\mu}_j \;,\; j=1,2,3,\dots,s \\[10pt] \varepsilon_s\bar{\bar{k}}^{-1}D s\dfrac{d\bar{\mu}_s}{r})_{r=a} + \bar{\mu}_s)_{r=a} \end{bmatrix}$$

$$\bar{\mu}_1(0) \text{ bded}$$

Call these eigenfunctions $\tilde{\bar{\mu}}_n$. Since L is self adjoint and positive definite we know that there is a set of orthonormal eigenvectors and a set of positive discrete eigenvalues and there is an expansion theorem for the functions $\bar{f}_j(r)$ belonging to a suitable Hilbert space so that

$$\tilde{\bar{f}}(r) = \begin{bmatrix} \bar{f}_j \\[10pt] \varepsilon_s\bar{\bar{k}}^{-1}\bar{\bar{D}}s\dfrac{d\bar{f}_s}{dr})_{r=a} + \bar{f}_s)_{r=a} \end{bmatrix} = \overset{\infty}{\underset{1}{\Sigma}}\omega_n\bar{\mu}_{jn}$$

where

$$\omega_n = (\tilde{\bar{f}}(r), \tilde{\bar{\mu}}_n)$$

and therefore a FFT may be defined to solve the system. By taking the inner product with the eigenvectors we obtain

$$\tilde{\bar{f}}(r) = \begin{bmatrix} \bar{f}_j \;,\; j=1,2,3,\dots,s \\[10pt] \varepsilon_s\bar{\bar{k}}^{-1}\bar{\bar{D}}s\dfrac{d\bar{f}_s}{dr})_{r=a} + \bar{f}_s)_{r=a} \end{bmatrix} = \overset{\infty}{\underset{1}{\Sigma}}\omega_n\tilde{\bar{\mu}}_n$$

where

$$\omega_n = (\bar{f}(r), \tilde{\bar{\mu}}_n)$$

and therefore a FFT may be defined by

$$F[\tilde{\bar{f}}(r), \tilde{\bar{\mu}}_n] = \omega_n$$

a scalar. Application to the original system gives

$$F[L\tilde{\bar{v}}_k] = -F[\dfrac{\partial}{\partial \theta}\bar{v}_k] + F[\tilde{\bar{g}}(\theta)] + \eta_n v_{kn}$$

$$= -\dfrac{\partial v_{kn}}{\partial \theta} + g_n(\theta)$$

and therefore

$$v_{kn}(\theta) = \int_0^\theta e^{-\eta_n(\theta-\tau)}g_n(\tau)d\tau$$

where we have assumed the initial values are all zero. Therefore

$$\tilde{\bar{v}}_k = \overset{\infty}{\underset{1}{\Sigma}}v_{kn}(\theta)\tilde{\bar{\mu}}_{kn}$$

where we have written the eigenfunction as $\tilde{\bar{\mu}}_{kn}$ since each of the radial eigenfunctions also depends upon k.

STEADY STATE IN A FIXED BED REACTOR

Consider now the same problem with the sphere but with homogeneous particles imbedded in a fixed bed in which axial dispersion plays a role. The model in dimensionless form is

$$\dfrac{\bar{\bar{D}}}{\rho^2}\dfrac{\partial}{\partial \rho}(\rho^2\dfrac{\partial \bar{\bar{c}}}{\partial \rho}) + \alpha\bar{\bar{K}}\bar{\bar{c}} = 0 \;,\; 0<\rho<1$$

$$-\bar{\bar{D}}\dfrac{\partial \bar{c}}{\partial \rho})_{\rho=1} = \bar{\bar{k}}(\bar{c}-\bar{C}) \;,\; \rho=1$$

$$\dfrac{\partial^2\bar{C}}{\partial s^2} - Pe\dfrac{\partial \bar{C}}{\partial s} - \gamma\bar{\bar{D}}\dfrac{\partial \bar{c}}{\partial \rho})_{\rho=1} = 0$$

$$\dfrac{\partial \bar{C}}{\partial s} = 0 \;,\; s=1$$

$$-\dfrac{\partial \bar{C}}{\partial s} + Pe\bar{C} = Pe\bar{C}_0 \;,\; s=0$$

We consider the same problem with n chemical species so that $\bar{\bar{D}}$ is a diagonal matrix and $\bar{K}$ is the Wei-prater matrix of kinetic coefficients. The Peclet number is a constant and the same for all species in the intersticial space. The usual exponential substitute $\bar{c}=\bar{v}\exp(Pes/2)$, $\bar{C}=\bar{V}\exp(e\,Pes/2)$ reduces the above to

$$\dfrac{1}{\rho^2}\dfrac{\partial}{\partial \rho}(\rho^2\dfrac{\partial \bar{v}}{\partial \rho}) + \alpha\bar{\bar{D}}^{-1}\bar{\bar{K}}\bar{v} = 0$$

$$-\bar{\bar{D}}\dfrac{\partial \bar{v}}{\partial \rho} = \bar{\bar{k}}(\bar{v}-\bar{V}) \;,\; \rho=1$$

$$\dfrac{\partial^2\bar{V}}{\partial s^2} - \dfrac{Pe^2}{4}\bar{V} - \gamma\bar{\bar{D}}\dfrac{\partial \bar{v}}{\partial \rho})_{\rho=1} = 0 \qquad (12)$$

$$\dfrac{\partial \bar{V}}{\partial s} + \dfrac{Pe}{2}\bar{V} = 0 \;,\; s=1$$

$$-\dfrac{\partial \bar{V}}{\partial s} + \dfrac{Pe}{2}\bar{V} = Pe\bar{C}_0 \;,\; s=0$$

We consider first the operator $\bar{\bar{\alpha}}\bar{\bar{D}}^{-1}\bar{\bar{K}}$ which is negative definite with respect to the inner product

$$<\bar{u},\bar{v}>_3 = \sum_{i=1}^{n} \frac{u_i v_i D_i}{a_i}.$$

Thus if we denote the eigenvalues by $-\lambda_k$ and eigenfunctions $\bar{z}_k$ application of the little fourier transform gives

$$\frac{1}{\rho^2}\frac{\partial}{\partial\rho}(\rho^2\frac{\partial v_k}{\partial\rho})-\lambda_k v_k=0$$

$$-\frac{\partial v_k}{\partial\rho}=S(v_k-V_k)$$

where we have assumed as earlier that $k_i/D_i=S$, a constant, that is $\bar{\bar{D}}^{-1}\bar{\bar{k}}=S\bar{\bar{I}}$. Thus

$$v_k=\frac{1}{\rho}\frac{SV_k\sinh\sqrt{\lambda_k}\rho}{(S-1)\sinh\sqrt{\lambda_k}+\sqrt{\lambda_k}\cosh\sqrt{\lambda_k}}=\tilde{v}_k V_k$$

and

$$\bar{v}=\sum_1^n\frac{\tilde{v}_k\bar{z}_k V_k}{<\bar{z}_k,\bar{z}_k>_3}=\sum_1^n\frac{\tilde{v}_k V_k\bar{z}_k}{<\bar{z}_k,\bar{z}_k>_3}.$$

Application of the little transform to Equation (11) gives

$$\frac{\partial^2 v_j}{\partial s^2}-\frac{Pe^2}{4}V_j-\gamma\sum_{k=1}^{n}\frac{(\frac{\partial v_k}{\partial\rho})\cdot<\bar{\bar{D}}\bar{z}_k,\bar{z}_j>_3}{<\bar{z}_k,\bar{z}_k>_3}V_k$$

which we rewrite as

$$\frac{\partial^2 v_j}{\partial s^2}-\frac{Pe^2}{4}V_j-\gamma\sum_{k=1}^{n}P_k q_{kj}V_k=0$$

where

$$q_{kj}=<\bar{\bar{D}}\bar{z}_k,\bar{z}_j>_3=q_{jk} \quad ; \quad P_k=(\frac{\partial v_k}{\partial\rho})_{\rho=1}<\bar{z}_k,\bar{z}_k>_3^{-1}$$

and thus

$$\frac{d^2U}{ds^2}=\frac{Pe^2}{4}\bar{U}+\gamma\bar{\bar{B}}\bar{U} \quad ; \quad \bar{U}^T=[V_1 V_2\ldots V_n]$$

where $\bar{\bar{B}}$ is a matrix with $P_j q_{ij}$ at position (i,j). If we define an inner product

$$<\bar{\mu},\bar{v}>_5=\sum_1^n P_j\mu_j v_j$$

then

$$<\bar{\bar{B}}\bar{\mu},\bar{v}>_5=<\bar{\mu},\bar{\bar{B}}\bar{\mu}>_5$$

so $\bar{\bar{B}}$ is a symmetric operator and its eigenvalues are real and

$$<\bar{\bar{B}}\bar{\mu},\bar{\mu}>_5=\sum_1^n\sum_1^n P_j P_i q_{ij}\mu_i\mu_j.$$

It can be shown that this quadratic form has a matrix whose principal minors are related to Gram determinants which are positive so that all of the eigenvalues $\{\omega_j\}$ of $\bar{\bar{B}}$ are positive.

Let the corresponding eigenvectors be $\{\bar{\omega}_j\}$. Applying the little fourier transform

$$<\bar{U},\bar{w}_k>_5=U_k$$

gives

$$\frac{d^2U_k}{ds^2}=\frac{Pe^2}{4}U_k+\gamma\omega_k U_k$$

which must be solved subject to

$$\frac{dU_k}{ds}+\frac{Pe}{2}U_k=0 \quad , \quad s=1$$

$$-\frac{dU_k}{ds}+\frac{Pe}{2}U_k=PeC_{ojk}$$

the solution of which is

$$U_k=PeC_{ojk}\frac{\alpha_k\cosh\alpha_k(1-s)+\frac{Pe}{2}\sinh\alpha_k(1-a)}{(\alpha_k^2+\frac{Pe^2}{4})\sinh\alpha_k+\alpha_k Pe\cosh\alpha_k}$$

where $\alpha_k^2=\frac{Pe^2}{4}+\gamma\omega_k$. Then

$$\bar{U}=\sum_1^n U_k\bar{w}_k$$

$$V_j=\sum_{k=1}^{n}U_k w_{jk}$$

where w_{jk} is the jth component of $\bar{w}_k$. Then

$$\bar{C}(s)=e^{\frac{Pes}{2}}\bar{V}=e^{\frac{Pes}{2}}\sum_{j=1}^{n}V_j\frac{\bar{z}_j}{<\bar{z}_j,\bar{z}_j>_3}$$

with the outlet concentration given by $\bar{C}(1)$ for each species.

ADDITIONAL COMMENT

In this presentation we have used the Wei-Prater kinetic matrix throughout although simpler kinetics can be handled just as well and normally more simply. If we consider instead of the completely coupled system the consecutive reversible scheme

$$A_1\overset{\rightarrow}{\leftarrow}A_2\overset{\rightarrow}{\leftarrow}A_3\overset{\rightarrow}{\leftarrow}\ldots\overset{\rightarrow}{\leftarrow}A_n$$

with rate constant for $A_i\rightarrow A_{i+1}$ as k_i and its reverse as rate k_i', then the reaction matrix is

$$\bar{\bar{K}}=\begin{bmatrix}-k_1 & -k_1' & 0 & - & - & - & 0\\ k_1 & -k_1'-k_2 & k_2' & - & - & - & 0\\ 0 & & & & & & \\ \vdots & & & & & & \vdots\\ \vdots & & & & & & \\ 0 & - & - & - & - & k_{n-1} & -k_{n-1}'\end{bmatrix}$$

This is a matrix with generic form

$$A = \begin{bmatrix} -\alpha_1 & \gamma_1 & 0 & - & - & -0 \\ \beta_2 & -\alpha_2 & \gamma_2 & - & - & -0 \\ 0 & & & & & \\ | & & & & & \\ | & & & & & \\ 0 & - & - & - & \beta_n & -\alpha_n \end{bmatrix}$$

a Jacobi matrix. For a variety of problems $\alpha_i > 0$, $\beta_i > 0$, $\gamma_i > 0$ and it can be shown that $\bar{\bar{A}}$ is symmetric on R^n with respect to the inner product.

$$(\bar{a}, \bar{b}) = \sum_1^n \delta_i a_i b_i$$

where

$$\delta_1 = 1 \; , \quad \delta_i = \frac{\gamma_1 \gamma_2 \cdots \gamma_{i-1}}{\beta_2 \beta_3 \cdots \beta_i} \; , \quad i = 2 \text{ to } n \; .$$

The problems in this paper can all be rephrased with these kinetics since

$$\delta_i = \frac{k_1' k_2' \cdots k_{i-1}'}{k_1 k_2 \cdots k_{i-1}}$$

and we see that δ_i is the product of equilibrium constants for each of the reactions and therefore is only temperature dependent. The inner product above can now be used throughout. It can be shown that the eigenvalues are all non-positive.

NOTATION

a	particle radius
A_i	ith chemical species
c, c_i, C_i	concentrations
$\bar{c}, \bar{C}$	concentration vectors
D	axial dispersion coefficient
$\bar{\bar{D}}, \bar{\bar{D}}^j$	intraparticle diffusion matrix
D^j	ith component diffusion matrix in jth shell
F	finite Fourier transform operator
$\bar{f}(r)$	initial concentration vector
f	little fourier transform operator
$G(\theta)$	time dependent ambient concentration
k_{ji}	reaction velocity constant $A_i \rightarrow A_j$
$\bar{\bar{K}}, \bar{\bar{K}}^j$	reaction matrix
$\bar{\bar{k}}$	masstransfer matrix
k_i	mass transfer coefficient for ith species
L	differential operator
N	number of particles per unit of volume
Pe	Peclet number
r	radius variable
s	number of shells in a particle
S	k_j / D_j

u	intersticial velocity
$\langle u, v \rangle_j$	inner product of jth kind
x	axial variable

Greek Symbols

α_j	surface factor to produce dimensionality
β	constant to produce dimensionality
γ	constant to produce dimensionality
$\varepsilon, \varepsilon_j$	porosity
λ	eigenvalue
ρ	dimensionless radius variable
θ	time
τ	time
ω	eigenvector

LITERATURE CITED

Aris, R., The Mathematical Theory of Diffusion and Reaction in Permeable Catalysts, Oxford Univ. Press, Oxford (1975).

Coddington, E. A. and Levinson, N., Theory of Ordinary Differential Equations, McGraw Hill Book Co., New York (1955).

Cole, Randall H., Theory of Ordinary Differential Equations, Appleton-Century-Crofts, New Yrok (1968).

Ramkrishna, D. and Amundson, Neal R., Chem. Eng. Sci., 28. 601 (1973)

Ramkrishna, D. and Amundson, Neal R., Chem. Eng. Sci., 29, 1457 (1974).

Ramkrishna, D. and Amundson, Neal R., J. Appl. Mech., 41, 1106 (1974a).

Ramkrishna, D. and Amundson, Neal R., Chem. Eng. Sci., 29, 1353 (1974).

Ramkrishna, D. and Amundson, Neal R., Linear Operators in Chemical Engineering with Applications to Transport and Reaction Systems, Prentice Hall Inc., Englewood Cliffs, NJ (1985)

Wei, James and Prater, C. D., "The Structure and Analysis of Complex Reaction Systems," in Advances in Catalysis, Academic Press, New York (1962).

THE CHANGING ROLE OF APPLIED MATHEMATICS IN CHEMICAL ENGINEERING

Stuart W. Churchill ■ Carl V.S. Patterson Professor of Chemical Engineering
University of Pennsylvania, Philadelphia, PA 19104

During the first-quarter century of the AIChE, from 1908 through 1932, our publications were almost wholly descriptive and experimental. Few invoked mathematics beyond calculus. The second-quarter century, running through 1957, saw a slow but steady increase in the use and complexity of mathematical models and methods of solution. In the just-past, third-quarter century, through 1982, this increase has continued. The fraction of papers in the *AIChE Journal* which invoke mathematics beyond the preparation of most undergraduates of the time has increased from about 20% in 1958 to about 40% in 1983. It is concluded that: (1) the development of digital computers and the related software has provided the principal impetus for the increased role of analyses; (2) the development of applied mathematics in chemical engineering has made an enormous and not fully credited contribution to the improvement, efficiency, safety and profitability of industrial operations; (3) the skill in applied mathematics of our post-graduate students in chemical engineering is a neglected national resource which. if called upon more fully, might contribute to a revival in the competitive international position of the chemical process industries; (4) the mathematical preparation of our undergraduates must be upgraded if the practitioners of our profession are to understand, utilize and encourage the utilization of the currently available techniques in mathematical modelling; and (5) the AIChE has continually decreased its relative commitment to the archival literature, and thereby to the development and dissemination of applied mathematics, throughout its history, and particularly so in recent times.

INTRODUCTION

To look backward and perhaps forward seems appropriate at this point in the life of our profession of chemical engineering--as demarked by the Diamond Jubilee Meeting of the American Institute of Chemical Engineers. Some understanding and guidance ought to be gained from an analysis of our roots, our history and our present status.

With this objective, an examination was undertaken of the mathematical content of our archival literature over the past 75 years, as exemplified by the Transactions of the American Institute of Chemical Engineers, as a separate publication from 1908 through 1946, as part of Chemical Engineering Progress in 1947, by Chemical Engineering Progress from 1948 through 1954, and by the AIChE Journal from 1955 through 1983. Papers in the Chemical Engineering Progress Symposium Series (later the AIChE Symposium Series), which was started in 1951, were arbitrarily omitted from this survey, although several issues in the 1960's were on applied mathematics, computing and statistics. The chosen sample encompasses only a fraction of the total archival literature of chemical engineering

University of Pennsylvania
Philadelphia, PA

and, regrettably, a rapidly decreasing fraction of the published work of our our membership. In any event, the selection of papers herein is probably representative of past if not present research in chemical engineering.

The exercise of scanning the archival literature of the AIChE over a seventy-five year period with respect to applications of mathematics and modelling proved to be somewhat overwhelming, but also quite informative and inspiring. Only with great self-discipline was attention confined to this particular subject. Tracing the thread of any topic through seventy-five years of our literature is highly recommended as a cultural as well as a practical pursuit.

An indication of the yearly contents of the chosen "archival" journals is provided in Table I. All of the papers in the Transactions and in Chemical Engineering Progress (1948-1954) are included in this listing, although some in every issue and most in the early years are nonarchival.

In order to hold this survey within reasonable limits, attention was arbitrarily confined to innovative papers, i.e., those which introduced particular mathematical techniques to the archival literature of the AIChE, and to those which invoked advanced mathematics, i.e., beyond that traditionally encompassed in the typical undergraduate curriculum in chemical engineering of each era. The identifications below of these pioneering applications are admittedly incomplete because a detailed reading of every paper to avoid overlooking some subtle inclusion was obviously not feasible. These identifications are also necessarily subjective, being based on a personal interpretation of what is innovative and what is advanced. Despite these limitations, the observed trends are probably representative.

As an invocation for this analysis, it may be appropriate to recall that Dr. Samuel P. Sandtler in the first Presidential address in December 1908 stated that one of the objects of the formation of the American Institute of Chemical Engineers was "to publish and distribute such papers as shall add to classified knowledge in chemical engineering and shall increase industrial activity." He went on to say that this   "is surely one of the most important of the whole list of aims to be accomplished"(1).

THE FIRST QUARTER CENTURY

The frontier work in our profession, as represented by our publications, was, as might be expected, initially descriptive and experimental. Over the first 25 years from 1908 to 1933 the papers which invoked mathematics beyond elementary calculus were remarkably few in number. For example, there were none such in Volume 1 of the Transactions (in 1908), and only one a quarter of a century later in Volume 30 (1933-34), which was composed primarily of papers presented at the Silver Anniversary Meeting. In the intervening 25 years, only 11 papers in all were identified by this survey as introducing a new and advanced application of mathematics in chemical engineering, and only 12 additional ones as requiring a mathematical background on the part of the reader beyond that normally required for the bachelors degree in chemical engineering at that time.

Examples of the few mathematically sophisticated papers during that first quarter century are by:

W.B. Van Arsdel (2) in 1921 on the theory of gas absorption.
J.H. Graff (3) in 1926 on statistical analysis.
G.G. Brown and C.C. Furnas (4) in 1926 on the solution of partial differential

equations by graphical integration.
H.C. Hottel (5) in 1929 on radiation from gases.
T.R. Running (6) in 1929 on graphical calculus.
    Professor Running who was a member of the Department of Mathematics at the University of Michigan, but was associated closely with the Department of Chemical and Metallurgical Engineering, was also scheduled to present a paper entitled "The Extrapolation of Experimental Data" at the Annual Meeting the following year, but it omitted and apparently never published because the subject was considered, in the absence of a preprint, "too abstruse a subject to satisfactorily present in the time left at our disposal" (7).
A.B. Newmann (8) in 1930 on the use of Bessel functions for thermal conduction in a cylinder.
T.B. Drew (9) in 1931 reviewing the theoretical solutions for laminar forced convection, some of which utilize expansions of eigenfunctions, similarity transformations and conformal mapping.
T.K. Sherwood (10) and A.B. Newman (11, 12) in 1931 on the drying of porous solids.
H.C.T. Eggers (13) in 1931 on smoothing data graphically, integrally and algebraically.

The number of pages of the Transactions gradually increased during this period (note that two volumes were published in 1928, 1929, 1930 and 1931), although not as rapidly as the membership. As a conclusion to this brief survey of our first quarter century, it may be noted that Fred C. Zeisberg, then Chairman of the Committee on Publication, reported at the Annual Meeting of 1932 that the number of pages had averaged 400 per year over the previous five years, and that the Transactions represented the largest single item of expenditure of the AIChE (14).

THE SECOND QUARTER CENTURY

The years from 1933 to 1958 saw a slow but steady increase in the breadth and complexity of mathematical models and solutions. The focus of the Golden Jubilee Meeting in 1958 was on the state of the art in various industries and fields. There were no specific sessions or papers on applied mathematics, and archival papers were not included in the Jubilee volume. However, one can infer from the titles of the papers in the program that some new and advanced applications of mathematics were included in the presentations, and some of these papers later appeared in the AIChE Journal.

In 1947 the Transactions were incorporated in, and in 1948 replaced by, a new journal, Chemical Engineering Progress, containing both archival and nonarchival material, and under the editorship of F.J. Antwerpen. Seven years later the archival papers were allocated to a separate publication, the AIChE Journal under the technical editorship of Harding Bliss, who was succeeded in November 1970 by Robert C. Reid, and in July 1976 by Robert H. Kadlec. The Transactions were monitored by an editorial board of prominent educators and practitioners. These changes in format and editorship do not appear to have produced any direct or abrupt transitions in the mathematical content and character of the archival papers.

Chilton (15) in reviewing the operation of Chemical Engineering Progress in an editorial in 1951 noted that at a cost of $4.50 per member, 95 technical articles were now being published per year as compared to an average of 37 per year in the Transactions from 1936-1945 at an inflation-adjusted cost of $2.75 per member out of 412 papers presented at AIChE meetings from September 1946-August 1950, 249 (or approximately two-thirds) were published, as well as 143 direct contributions. Publication and individual sale of the Chemical Engineering Progress Symposium Series was planned to begin later that year to relieve pressure on the pages of Chemical Engineering Progress.

By 1958, the papers utilizing mathematics beyond that studied by undergraduates had increased to 18 out of 91 or approximately 20%. Two major developments appear to be responsible for this latter significant percentage. First, courses in applied mathematics, supported by the pioneering textbooks of Sherwood and Reed (16) in 1937, and Marshall and Pigford (17) in 1947, began to appear in the curriculum of chemical engineering, at least at the graduate level. Second, this period witnessed the appearance of first analog and then digital computers. The rapid improvement in digital hardware and software soon resulted in the complete displacement of the analog devices. Some spokesmen for our profession openly deplored and criticized both of these developments. Sophistication in mathematical modeling and methods of solution was asserted to divert

interest and attention from problems that needed to be solved in favor of those that could be modeled and solved. On the other hand, some of our mathematically inclined members scorned and deprecated the solution of mathematical models with a computer as inelegant and unfair. Time has blunted the latter if not the former criticism.

It is more difficult to assess credit and priority for the developments in applied mathematics during these second twenty-five years, particularly since many of the pioneering papers, even by our own members, began to appear in non-AIChE publications, primarily because of a severe limitation on the number of archival pages. The following papers from the Transactions of the American Institute of Chemical Engineers, Chemical Engineering Progress (1948-1954 only), and the AIChE Journal are therefore offered only as representative examples.

Colburn (18) in 1938 on the manipulation of equations for condensation.
Brown and Souders (19) in 1933 on rearrangement of the equations for distillation.
Higbie (20) in 1934 on transient absorption.
MacMullen and Weber (21) in 1934 on short-circuiting in stirred reactors.
Johnstone, Pigford and Chapin (22) in 1941 on the application of exponential integrals and Bessel functions.
Tiller and Tour (23) in 1944 on applications of the calculus of finite differences to stagewise operations.
Grossman (24) in 1946 on graphical finite-difference calculation of the transient temperature distribution in catalytic convertors. This was the first of many graphical solutions of such partial differential equations, but such methods were soon to be made obsolete by digital computers.
Amundson (25) in 1946 on the application of matrices and finite-difference equations to binary distillation.
Wilhelm, Johnson, Wynkoop and Collier (26) on the use of electrical networks to solve partial differential equations. This technique also soon became obsolete owing to the rapid development of digital computation.
Douglas and Peaceman (27) in 1955, in the first volume of the AIChE Journal, proposing the still widely used ADI (alternating-direction-implicit) method

of finite-difference calculations.
Bilous and Amundsen (28), in the succeeding paper in the same issue, on the use of the analog computer and the Laplace transform for the analysis of reactor stability.
Said (29) in 1956 and (30) in 1958 on Poisson functions for the analysis of chromatographic columns.
Friedlander (31) in 1957 on asymptotic solutions for unconfined convection.
Edmister (32)in 1957 on digital computer solutions for distillation.
Churchill (33) in 1957 on approximate Laplace inversions.
Acrivos (34) in 1958 on integral boundary layer solutions for convection.

The count in Table 1 of papers introducing mathematical concepts or requiring a knowledge of advanced mathematics is discontinued in 1955 with the beginning of the AIChE Journal, owing to their rapid increase in number and the increasing difficulty of defining and identifying such.

THE THIRD QUARTER CENTURY

The third quarter century of the AIChE has now passed into history. Of the papers in the last 6 issues of AIChE Journal through September 1983, 67 out of 169 or approximately 40% invoke mathematical aspects that are not known to most of our undergraduates. The number and variety of such papers in the past 25 years makes their categorization difficult. Now, only a small fraction of the archival and pioneering papers are published in our own literature, partly because of the extensive development of chemical engineering throughout the world, but primarily because of the continued constraint on our archival pages.

The following papers from the AIChE Journal are therefore only intended to be representative of pioneering work on the indicated topics during the first half of the past quarter century:

1. Asymptotics--Acrivos (35), Hamill and Bankoff (36), Churchill (37), Churchill and Usagi (38).
2. Optimization--Lapidus et al. (37), Rudd et al. (40), Wilde (11), Lee (42), Aris et al. (43).
3. Finite-difference solutions for convection--Brian (44), Hellums and Churchill (45), Stone and Brian (46), Fussell and Hellums (47).

4. Stagewise calculations--Murdoch (48), Lemlich and Leonard (49), Martin (50).
5. Method of weighted residuals and variational principles--Schechter (51), Stewart (52), Sparrow (53), Sani (54), Wasserman and Slattery (55), Snyder et al. (56), Denn and Aris (57), Finlayson and Scriven (58).
6. Analysis of reacting systems--Beutler (59), Snow et al. (60), Wei and Prater (61), Zeman and Amundson (62), Bischoff (63).
7. Process dynamics and stability--Ellington and Ceaglske (64), Hsu and Gilbert (65), Leathrum et al. (66), Wilde (67), Amundson and Raymond (68), Gura and Perlmutter (69), Blum (70).
8. Solutions for multicomponent diffusion--Cussler and Lightfoot (71), Toor (72).
9. The derivation of similarity transformations--Hellums and Churchill (73).
10. Monte Carlo calculations--Howell and Perlmutter (74).
11. Computer-aided design--Lee and Rudd (75).
12. Analytical methods for convection--Larkin (76), Slattery (77), Farrell and Leonard (78), Snyder (79).
13. Statistical evaluation of constants--Schiffe (80).

CONCLUSIONS

What can one conclude from these numbers and abstractions?

A continuous increase in the use and sophistication of mathematical models over the history of our profession is indisputable. There may be some who are still skeptical of the overall contribution of this body of work to the design, operation, safety and profitability of chemical plants, but they are surely short-sighted. True, some excesses have occurred, namely: (1) the development of solutions for models or conditions that do not have value even as illustrations or bounds; and (2) the narrow education of some graduate students who neglect all other aspects of chemical engineering in favor of some branch of analysis. The latter graduates are as a consequence of their narrow preparation, ill-prepared to utilize their knowledge or even to evaluate their own work. If they turn to teaching, as they do disproportionately, they may propagate this deficiency.

Scanning the archival articles published by our professional society during the past seventy-five years with special attention to applications of mathematics is an edifying experience. Such a trek through our history affords a different impression of many papers than that obtained at the time they were published. It becomes obvious that most of the advances in processes and equipment, and in design, operation and safety, as well as in insight and understanding, have been signalled or directed by the more mathematically oriented papers--some of which were considered esoteric or irrelevant at the time of their publication. Most new products result from experimentation rather than from analysis. However, their successful transition into products often depends critically on analysis and modeling. The real economic value of publications is difficult to assess or appreciate because of their diverse and delayed utilization. Even direct applications may only be the "tip of the iceberg." Their real impact may be through changes in the "mind set" of the readers and, in the case of teachers, of their students. The influence of improved equipment, new processes or added data is usually quite obvious. The influence of mathematical concepts, models, methods of solution, and even solutions themselves is more subtle. A truly radical idea is difficult to understand and to accept. However, after finally being comprehended it is frequently accepted as second-nature or obvious.

Analyses such as those cited above have undoubtedly had a greater positive impact on industrial operations than is generally recognized. Our reactors, distillation columns, heat exchangers, fermentors, dialysis machines, etc., are designed with ever greater precision and confidence, and operate ever more efficiently, dependably and safely because of these sophisticated analyses. Indeed, the future of the chemical process industry may now depend on its ability to utilize such techniques more rapidly and more fully. Whether the chemical process and petroleum industries mature and decay, or show renewed growth and international leadership may depend on whether or not, and how fully, they commit themselves to the associated research and innovation. If the chemical and petroleum industries, which are currently in a state of retrenchment, are to flourish

economically, it may well be because of the innovations and cost-savings which result from fundamental research and analysis.

The skill in mathematical modeling and analysis which exists among their chemical engineering employees, particularly those with graduate training, is an under-utilized resource of the chemical process industries in our country. The recognition and perceptive use of this resource could be a favorable factor for renewed growth and prosperity. Furthermore, as noted below, this skill could easily be improved for those who terminate with a bachelors degree if it were in greater demand.

Formulating models which truly represent the physical world and are in a sound mathematical form is one of the higher callings of the modern chemical engineer. We have in the past been deeply indebted to mathematicians for the classical methods of analysis. The digital computer has now liberated us somewhat from this dependence, but has imposed new demands. Computer methods for chemical engineers must increasingly be developed by chemical engineers and related professionals rather than by mathematicians or "computer specialists" because special-purpose methods, often based on heuristics, are usually more efficient than general methods. Elements of pure mathematics will, however, continue to be helpful in organizing and simplifying our models.

Despite the unmistakable growth in the extent and depth of mathematical modeling and analysis, as documented by our archival literature, these skills have not been fully inculcated in our undergraduates. The preparation in mathematics in our secondary schools, at least for the superior students who elect chemical engineering, is better than at any time in the past. However, their mathematical education during the bachelors program has not advanced commensurately. Greater improvement has occurred at the graduate level. In the other developed countries, such as France, Great Britain, Germany and Japan, the mathematical preparation at all levels is superior to ours. What leadership we may have in modeling has been accomplished in spite of this handicap.

The most serious consequence of the outdated mathematical preparation of our undergraduates may be in the resulting incomprehensibility of much of the archival literature. By 1958, a B.S. graduate was unprepared to read perhaps 20% of the then current literature. By 1983 this inaccessible fraction had increased to as much as 40%. Most graduate students do develop an adequate capability in applied mathematics and the confidence to use it. Our educators thus have an obligation to improve the preparation of our undergraduates so that they may have access to the results of the research and analysis in their own field. Otherwise, those who develop new mathematical models and application will be isolated intellectually from the majority of practitioners, and the results of their analyses will go unrecognized and unused.

Because of the intellectual tradition we inherited from the chemists, chemical engineers have been more devoted to the development of new models and methods, and more receptive to their practical implementation than in other fields of engineering. Also, the commonality of our profession has been notably higher. On this 75th anniversary of our professional society we are in danger of losing these distinctions. Recognition of the long-range contribution of analysis by industry is overdue, as is preparation of our undergraduates to read the literature of analysis and innovation.

Vol.	Year(s)	Pages	Number of Articles			Members
			Total	Innov. Math.[1]	Adv. Math.[2]	
34	1938	670	36	0	0	1733
35	1939	720	42	0	0	1984
36	1940	304	48	0	1	2255
37	1941	959	45	1	1	2527
38	1942	1022	51	0	1	2923
39	1943	851	34	0	0	3347
40	1944	772	43	1	1	3929
41	1945	803	40	0	0	4866
42	1946	1020	66	2	2	5769

Chemical Engineering Progress[3]

43	1947	730	70	1	2	6727
44	1948	946	115	1	4	6914
45	1949	754	89	0	1	7914
46	1950	642	88	0	3	8938
47	1951	730	98	2	2	9677
48	1952	649	89	6	3	10411
49	1953	621	99	3	0	11428
50	1954	632	105	4	2	12588

1-Innovative use of applied mathematics in that journal.

2-Invoking mathematics beyond that required for the B.S. degree in chemical engineering.

3-A page in Chemical Engineering Progress averaged almost three times as many words a page as in the Transactions.

LITERATURE CITED

1. Sandtler, S.P., Trans. AIChE, 1, 35 (1908)
2. Von Arsdel, W.B., Trans. AIChE, 14, 391 (1921-1922).
3. Graff, J.H., Trans. AIChE, 18, 165 (1926)
4. Brown, G.G. and C.C. Furnas, Trans. AIChE 18, 295 (1926).
5. Hottel, H.C., Trans. AIChE, 19, 173 (1927).
6. Running, T.R., Trans. AIChE, 23, 159 (1929).
7. White, A.H., Trans. AIChE, 24, 238 (1930).
8. Newman, A.B., Trans. AIChE, 26, 44 (1930).
9. Drew, T.B., Trans. AIChE, 26, 26 (1931).
10. Sherwood, T.K., Trans. AIChE, 27, 190 (1931).
11. Newman, A.B., Trans. AIChE, 27, 203 (1931).
12. Newman, A.B., Trans. AIChE, 27, 310 (1931).
13. Eggers, H.C.T., Trans. AIChE, 27, 334 (1931).
14. Zeisberg, F.C., Trans. AIChE, 28, 319 (1932).
15. Chilton, T.H., Chem. Eng. Prog., 47, 1 (1951).
16. Sherwood, T.K., and C.E. Reed, Applied Mathematics in Chemical Engineering, McGraw-Hill, New York (1939).
17. Marshall, W.R., Jr., and R.L. Pigford, The Application of Differential Equations to Chemical Engineering Problems, University of Delaware, Newark, DE (1947).

TABLE I

Publications in Applied Mathematics

Transactions of the American Institute of Chemical Engineers

Vol.	Year(s)	Pages	Number of Articles			Members
			Total	Innov. Math.[1]	Adv. Math.[2]	
1	1908	150	10	0	0	40
2	1909	239	19	0	0	101
3	1910	302	26	0	0	181
4	1911	417	31	0	0	145
5	1912	254	29	0	0	170
6	1913	238	19	0	0	195
7	1914	279	19	0	0	214
8	1915	278	19	0	0	220
9	1916	392	18	0	1	230
10	1917	438	21	0	0	270
11	1918	382	31	0	0	283
12	1919	531	32	0	0	306
13	1920	782	41	0	0	344
14	1921-22	448	31	1	0	454,529
15	1923	687	36	0	0	565
16	1924	495	33	0	0	613
17	1925	262	12	0	0	644
18	1926	472	31	2	1	681
19	1927	234	17	1	1	723
20	1928	272	15	0	0	788
21	1929	138	9	0	1	
22	1929	221	13	1	0	805
23	1929	191	14	0	0	
24	1930	239	11	1	2	872
25	1930	249	14	0	0	
26	1931	262	13	1	2	944
27	1931	401	21	4	4	
28	1932	283	26	0	0	1017
29	1933	323	11	0	1	1066
30	1933-34	625	29	1	2	1099
31	1934-35	708	37	1	2	1157
32	1936	483	28	0	1	1321
33	1937	542	31	0	0	1435

18. Colburn, A.P., Trans. AIChE, 30, 187 (1933-34).
19. Brown, G.G. and Mott Souders, Jr., Trans. AIChE, 30, 438 (1933-34).
20. Higbie, Ralph, Trans. AIChE, 31, 365 (1934-35).
21. MacMullin, R.B., and M. Weber, Jr., Trans. AIChE, 31, 409 (1934-1935).
22. Johnstone, H.F., R.L. Pigford and J.H. Chapin, Trans. AIChE, 37, 95 (1941).
23. Tiller, F.M., and R.S. Tour, Trans. AIChE, 40, 317 (1944).
24. Grossman, L.M., Trans. AIChE, 42, 535 (1946).
25. Amundson, Neal, Trans. AIChE, 42, 939 (1946).
26. Wilhelm, R.H., W.C. Johnson, R. Wynkoop and D.W. Collier, Chem. Eng. Prog., 44, 105 (1948).
27. Douglas, Jim, Jr., and D.W. Peaceman, AIChE J., 1, 505 (1955).
28. Bilous, Olegh, and N.R. Amundson, AIChE J., 1, 513 (1955).
29. Said, A.S., AIChE J., 2, 477 (1956).
30. Said, A.S., AIChE J., 4, 290 (1958).
31. Friedlander, S.K., AIChE, 3, 43 (1957).
32. Edmister, W.C., AIChE J., 3, 165 (1957).
33. Churchill, S.W., AIChE J., 3, 389 (1957).
34. Acrivos, Andreas, AIChE J., 4, 285 (1958).
35. Acrivos, Andreas, AIChE J., 6, 410 (1960).
36. Hamill, T.D. and S.G. Bankoff, AIChE J., 9, 741 (1963).
37. Churchill, S.W., AIChE J., 4, 431 (1965).
38. Churchill, S.W. and R. Usagi, AIChE J., 18, 1172 (1972).
39. Lapidus, Leon, Eugene Shapiro, Saul Shapiro and R.E. Stillman, AIChE J., 7, 288 (1961).
40. Rudd, D.F., Rutherford Aris and N.R. Amundson, AIChE J., 7, 376 (1961).
41. Wilde, D.J., AIChE, 9, 186 (1963).
42. Lee, E.S., AIChE J., 10, 309 (1964).
43. Aris, Rutherford, G.L. Nemhauser and D.J. Wilde, AIChE J., 10, 913 (1964).
44. Brian, P.L.T., AIChE J., 7, 367 (1961).
45. Hellums, J.D. and S.W. Churchill, AIChE J., 8, 690, 692, 719 (1962).
46. Stone, H.L. and P.L.T. Brian, AIChE J., 9, 681 (1963).
47. Fussell, D.D. and J.D. Hellums, AIChE J. 11, 833 (1965).
48. Murdoch, P.G., AIChE J., 7, 526 (1961).
49. Lemlich, Robert and R.A. Leonard, AIChE J., 8, 214 (1962).

50. Martin, J.J., AIChE J., 9, 646 (1963).
51. Schechter, R.S., AIChE J., 7, 445 (1961).
52. Stewart, W.E., AIChE J., 8, 425 (1962).
53. Sparrow, E.M., AIChE J., 8, 599 (1962).
54. Sani, R.L. AIChE J., 9, 279 (1963).
55. Wasserman, M.L. and J.C. Slattery, AIChE J., 10, 383 (1964).
56. Snyder, L.J., T.W. Spriggs and W.E. Stewart, AIChE J., 10, 535 (1964).
57. Denn, M.M. and Rutherford Aris, AIChE J., 11, 367 (1965).
58. Finlayson, B.A. and L.E. Scriven, AIChE J., 12, 1151 (1966).
59. Beutler, J.A., Chem. Eng. Prog., 50, 569 (1954).
60. Snow, R.H. R.E. Peck and C.G. Von Fredersdorf, AIChE J., 5, 305 (1959).
61. Wei, James and C.D. Prater, AIChE J., 9, 77 (1963).
62. Zeman, Ronald and N.R. Amundson, AIChE J. 9, 297 (1963).
63. Bischoff, K.B., AIChE J., 11, 351 (1965).
64. Ellington, W.R. and N.H. Ceaglske, AIChE J., 5, 30 (1959).
65. Hsu, J.P. and Nathan Gilbert, AIChE J., 8, 593 (1962).
66. Leathrum, J.F., E.F. Johnson and Leon Lapidus, AIChE J., 10, 16 (1964).
67. Wilde, D.J., AIChE J., 11, 237 (1965).
68. Amundson, N.R. and L.R. Raymond, AIChE J. 11, 339 (1965).
69. Gura, I.A. and D.D. Perlmutter, AIChE J., 11, 475 (1965).
70. Blum, E.H., AIChE J., 11, 532 (1965).
71. Cussler, E.L., Jr. and E.N. Lightfoot, Jr., AIChE J., 9, 702 (1963).
72. Toor, H.L., AIChE J., 10, 448 (1964).
73. Hellums, J.D. and S.W. Churchill, AIChE J., 10, 110 (1964).
74. Howell, J.R. and Morris Perlmutter, AIChE J., 10, 562 (1964).
75. Lee, W.Y. and D.F. Rudd, AIChE J., 12, 1184 (1966).
76. Larkin, B.K. AIChE J., 7, 530 (1961).
77. Slattery, J.C., AIChE J., 5, 663 (1962).
78. Farrell, M.A. and E.F. Leonard, AIChE J., 9, 190 (1963).
79. Snyder, Wm. F., AIChE J., 9, 503 (1963).
80. Schiffe, Henry, Chem. Eng. Prog., 50, 200 (1954).

URANIUM ENRICHMENT— PAST, PRESENT AND FUTURE

Manson Benedict ■ Nuclear Engineering Department, Massachusetts Institute of Cambridge, MA

The world's first uranium enrichment plants, built at Oak Ridge, TN, by the Manhattan Project in 1944-1946, were the Y-12 Electromagnetic Plant, The S-50 Thermal Diffusion Plant, and the K-25 and K-27 Gaseous Diffusion Plants. Also, gas centrifuge enrichment pilot plants were built and run by the University of Virginia and the Standard Oil Company of New Jersey.

In the 1950's, larger gaseous diffusion plants were built at Oak Ridge, Paducah, KY, and Portsmouth, OH, and in the Soviet Union, England, France and China.

Two large enrichment plants that started operation in the 1970's were the Eurodif gaseous plants in France and the Urenco-Centec gas centrifuge plants in England, the Netherlands and West Germany.

Enrichment plants now being built include aerodynamic process separation pilot plants in Brazil and South Africa, gas centrifuge plants in the United States and Japan, and a laser separation pilot plant in the United States.

These projects are briefly described.

This paper describes the beginnings of uranium enrichment technology, its present status, and its future possibilities.

1. BEGINNINGS

In World War II the four processes listed in Figure 1 were developed by the Manhattan Project.

PROCESS	PRODUCT	DISPOSITION
ELECTROMAGNETIC (CALUTRON)	Y-12 PLANT, 90% U-235	MADE FIRST U-235, SHUT DOWN AFTER K-25 FULLY OPERATIONAL
THERMAL DIFFUSION	S-50 PLANT, 0.86% U-235	MADE ENRICHED FEED FOR Y-12
CENTRIFUGE	MACHINE, CAPACITY 1 SWU YR	PILOT PLANT ONLY
GASEOUS DIFFUSION	K-25, K-27 PLANTS, 90% U-235	OPERATED UNTIL 1964. K-29, K-31 AND K-33 BUILT IN 1950s

Figure 1. Uranium enrichment processes developed by U.S. Manhattan Project, 1943-1946.

1.1 Electromagnetic Process

The electromagnetic process was conceived by Ernest Lawrence, who adapted his cyclotron at the University of California to separate uranium-235 from

Massachusetts Institute of Technology, Cambridge, Massachusetts.

natural uranium. By February 1942 Lawrence had produced 225 micrograms of uranium enriched to 30% U-235. Thus encouraged, General Groves directed Stone and Webster to design and build the Y-12 plant to produce hundreds of grams of 90%-enriched U-235 per day. The first separating unit, called an alpha calutron, operated in September 1943. Since it enriched uranium-235 to only 15% a second type, called a beta calutron, was needed for enrichment to 90%.

An alpha calutron is shown in Figure 2. It consisted of a large electromagnet between whose pole pieces were set 96 vacuum

Figure 2. Electromagnetic process—Alpha units at Y-12 plant.

tanks. In each tank uranium tetrachoride was vaporized, dissociated, and the uranium metal ionized. The U-235 and U-238 ions were accelerated electrostatically and focussed in slightly separate orbits by the magnet. Only a tenth of the feed was separated. The remaining uranium, smeared all over the vacuum tank, had to be recovered periodically by chemical leaching.

Over a year was required to solve the electrical, mechanical and vacuum problems of the calutrons. But by November 1944 operation by Tennessee Eastman was sufficiently reliable to permit Groves to authorize construction of nine alpha units and six betas. The Y-12 plant in 1947 is shown in Figure 3.

Figure 3. Y-12 Electromagnetic plant, 1947.

Costs were high. 13,000 operators were needed to run the calutrons, and 10,000 more were used for the messy batch chemistry to recover uranium from them. But the process worked and produced most of the U-235 made during the war. However, when the less labor-intensive gaseous diffusion plant was completed, in 1946 the electromagnetic process shut down.

1.2   Thermal Diffusion

In 1940 at the Bureau of Standards Philip Abelson found that uranium isotopes could be partially separated by thermal diffusion of UF_6 at supercritical conditions. Larger scale development was continued at the Naval Research Laboratory. In June 1944 Groves directed that the S-50 thermal diffusion plant be built near the plant later to supply steam for the gaseous diffusion plant. A plant with 2142 thermal diffusion columns 48 feet high was built by the H.K. Ferguson Co. in just 90 days. It went into full operation in March 1945.

Although very inefficient and a veritable inferno of leaking steam, it did enrich uranium-235 to 0.89% which, fed to the Y-12 plant, there advanced production of 90% U-235 by a week. In September 1945, when the more efficient gaseous diffusion plant needed its steam, the S-50 plant shut down.

Figure 4 is an interior view and Figure 5 an exterior view of the S-50 plant.

Figure 4. Thermal diffusion process (1945).

Figure 5. Thermal diffusion process—S-50 Plant.

1.3.   Gas Centrifuge

In 1941 Jesse Beams at University of Virginia partially separated uranium isotopes by whirling UF_6 in a small high-speed gas centrifuge. Harold Urey suggested that separation be enhanced by countercurrent flow, for which Karl Cohen

developed relevant theory. In 1942
Westinghouse manufactured longer centrifuges
tested by Beams and by Eger Murphree of
Standard Oil Company of New Jersey.

One Duralumin machine 345 centimeters
long had a capacity of 0.94 kilograms
separative work per year at a peripheral
speed of 205 meters per second. Another
faster machine one meter long shown in
Figure 6 had a capacity of 0.8 kilograms per
year and produced 200 pounds of UF$_6$ enriched
five percent in U-235 between October 1943
and January 1944. Then, because gaseous
diffusion was judged more practical, work on
the centrifuge was halted.

Figure 6. Centrifuge under test at University of Virginia.

The Beams centrifuge was a complex,
four-stream machine, with oil-lubricated
bearings and gas-buffer seals. Ten years
later with the invention of a simpler,
three-stream machine, Gernot Zippe, an
Austrian engineer, made the centrifuge again
a serious competitor of gaseous diffusion.

1.4 Gaseous Diffusion

In 1940 George Kistiakowsky and Urey
suggested that gaseous diffusion might be
used to concentrate U-235. Studies and
laboratory work on the process were
undertaken in Urey's laboratory at Columbia
University under the dynamic leadership of
John Dunning.

The problems were formidable. A
diffusion barrier with billions of holes
less than a tenth of a micrometer in
diameter would be needed. The barrier and
the balance of plant would have to withstand

corrosion by uranium hexafluoride, the only
known volatile uranium compound. Special
pumps, seals and valves would have to be
developed.

In January 1942 the OSRD asked P.C.
(Dobie) Keith of the M.W. Kellogg Co. to see
how these problems might be solved and to
study design of a plant to produce one
kilogram of 90% enriched uranium per day.
These studies were so encouraging that in
December Groves decided that a gaseous
diffusion plant should be built in addition
to the electrogmagnetic plant.

A Kellogg subsidiary, Kellex
Corporation, with Keith, Technical Director,
and Albert Baker, Project Manager, was asked
to complete development and engineer the
plant and oversee construction. J.A. Jones
Construction Co., under Edwin Jones, was
asked to build the plant, named K-25, and
Union Carbide, under George Felbeck, was
named operator.

In the first three months of 1943
Kellex worked out the process design of the
plant. Over 4000 diffusion stages would be
needed, each with a novel compressor for UF$_6$
and a diffusion barrier not then in
existence. To prevent loss of UF$_6$, the
plant would have to be leak tight to an
unprecedented degree, and all plant surfaces
contacting UF$_6$ would have to be
corrosion-resistant nickel or copper.

During 1944 difficult development
problems were solved. Allis Chalmers built
a new factory to make the special
compressors. Judson Swearingen developed a
novel seal which minimized leakage of air
and UF$_6$. Du Pont and Harshaw made the
fluorine and perfluoroxylene coolant in
requisite amounts. Mass spectrometers for
leak detection and process analysis designed
by Alfred Nier were made by General
Electric. Novel vacuum-tight valves
designed by J.C. Hobbs were built by Crane.
Chrysler started to make coolers and
diffusers.

But diffusion barrier of adequate
strength and separation was still not
available. At the last possible moment,
desperate cooperative research by Dunning's
laboratory, Kellex and Carbide developed a
suitable barrier, which was manufactured by
Houdaille Hershey late in 1944.

Meanwhile, 20,000 construction workers
were building the K-25 diffusion plant at

Oak Ridge. The first stages ran on UF$_6$ in January 1945. As additional stages were connected, the U-235 concentration slowly increased. By August 1945 the last K-25 stages started operation and their U-235 content was high enough to permit shutdown of Y-12's alpha stages. By 1946 K-25 was making 90% U-235 and Y-12's beta stages were shut down.

Figure 7 shows two stages of the K-25 plant. Figure 8 is an aerial view of its 4000-stage cascade.

In the next five years diffusion plant capacity was increased by construction of the K-27 and K-29 plants, with progressively larger equipment than K-25.

Figure 7. Interior of K-25 gaseous diffusion plant.

Figure 8. Aerial view of K-25 plant.

2. PRESENT STATUS

2.1 Present Producers of Separative Work

The countries presently producing separative work are listed in Figure 9.

COUNTRY	LOCATION	PROCESS	CAPACITY MSWU/YR
U.S.A.	Oak Ridge, TN Paducah, KY Portsmouth, OH	Gaseous Diffusion	27
USSR	?	Gaseous Diffusion	7-10
ENGLAND	Capenhurst	Gaseous Diffusion	0.4
FRANCE	Pierrelatte	Gaseous Diffusion	0.5
CHINA	Lanchow	Gaseous Diffusion	0.2
FRANCE BELGIUM ITALY ET AL.	Tricastin (Eurodif)	Gaseous Diffusion	10.8
ENGLAND HOLLAND W. GERMANY	Capenhurst Almelo	Gas Centrifuge	0.9

Figure 9. Present-day uranium enrichment plants.

After the successful use of gaseous diffusion in the United States, the Soviet Union, England, France and China, in that order, built and operated gaseous diffusion plants to make fully enriched uranium, primarily for military purposes. In addition, up to 3 million SWU per year of Soviet low-enriched uranium has been sold abroad for civilian power plants. The Eurodif gaseous diffusion plant, owned by a consortium of French, Belgian, Italian, Spanish and Iranian interests, makes low-enriched uranium for civilian purposes. Urenco and Centec, two associated consortia of English, Dutch and West German organizations, have developed an improved gas centrifuge and are selling low-enriched uranium.

2.2 U.S. Plants

Today's U.S. diffusion plants are listed in Figure 10. In the early 1950's the capacity of the Oak Ridge diffusion plant was increased to 4.73 million separative work units (SWU) per year by

YEAR	LOCATION	FULLY POWERED SEPARATIVE CAPACITY, MILLION SWU/YR
1945-54	OAK RIDGE	4.73
1953-54	PADUCAH	7.31
1955-56	PORTSMOUTH	5.19
	CAPACITY EXPANSION PROGRAMS	
1973-81	CASCADE IMPROVEMENT	5.50
1975-82	CASCADE UPRATING	4.60
TOTAL		27.33

Figure 10. Schedule of completion of U.S. gaseous diffusion facilities.

installation of larger stages, incorporating major improvements developed by Union Carbide. At the same time Carbide built a second, 7.31 million SWU/year diffusion plant at Paducah, Kentucky. This consisted entirely of large stages and was used for producing slightly enriched uranium. In 1955-56 a third, 5.19 MSWU/yr diffusion plant was built at Portsmouth, Ohio, for operation by Goodyear Atomic Co. This plant has over 4000 stages and can produce fully enriched uranium.

Figure 11 shows large diffusion stages

Figure 11. Gaseous diffusion cell interior.

inside the Oak Ridge plant. Figures 12, 13 and 14 are aerial views of today's diffusion plants at Oak Ridge, Paducah and Portsmouth.

Figure 12. Oak Ridge gaseous diffusion plant.

In the past ten years Carbide devised the means for expanding diffusion plant capacity listed at the bottom of Figure 10. Cascade improvement consisted in substituting more efficient barrier and compressors and reducing pressure losses. Cascade uprating consisted in providing larger motors and cooling capacity and increasing electric power capability from 6100 to 7400 megawatts. Today, because of

Figure 13. Paducah gaseous diffusion plant.

Figure 14. Portsmouth gaseous diffusion plant.

present low demand for separative work, U.S. plants operate at reduced pressure and power levels, using 2500 megawatts to produce ten million separative work units per year.

The price charged for separative work from U.S. plants has increased from $26 per SWU in 1967 to about $140 per SWU today. Most of the change is due to the large increase in the cost of electricity. Around 2200 kilowatt hours are needed to produce one SWU by gaseous diffusion.

2.3 Eurodif Diffusion Plant

An aerial view of the 1400-stage Eurodif plant is shown in Figure 15. A feature of the plant is its use of four 936-megawatt pressurized water reactors to provide dedicated electric power. This protects the Eurodif project from the power cost increases which have eroded the competitive cost advantage of U.S. diffusion plants. Another feature is Eurodif's

Figure 15. Eurodif gaseous diffusion plant.

modular stage construction, which mounts compressor, cooler and diffuser in a single unit.

Design was completed and construction started in 1973. UF$_6$ was charged to the first stages in 1978, and the first slightly enriched product was withdrawn in February 1979. As stages were completed, plant capacity increased from 2.5 million SWU per year in September 1979 to its present full capacity of 10.8 million.

2.4 Urenco-Centec Centrifuge Plants

Although gas centrifuges will likely supply much of the world's future separative work, only one group of centrifuge plants is now in production, that of the English, Dutch, West German Urenco-Centec organization, with plants operating in Capenhurst, England and Almelo, Holland and a future plant planned in Germany. Construction of three 20-ton-SWU-per-year pilot plants started in 1971, and full production was reached in 1976. Construction of two commercial plants, at Capenhurst and Almelo, was started in 1975,

Figure 16. Erenco Centrifuges.

with production commencing in 1977. Present capacity is 900 tons per year. Expansion to 2130 tons per year is scheduled for completion in the late 1980's. Further expansion to 10,000 tons per year is possible.

Figure 16 is an interior view of a Urenco plant in Almelo.

Capacity of individual Urenco centrifuges is thought to be between five and ten SWU/year; they are designed for a mean life of eight years or more.

3. FUTURE PROJECTS

3.1 Principal Projects

Figure 17 lists additional enrichment projects likely to produce in the future.

COUNTRY	PROCESS	STATUS
U.S.A.	GAS CENTRIFUGE	2.2 MSWU/YR PLANT UNDER CONSTRUCTION. EXPANSION TO OVER 12 MSWU/YR PLANNED
JAPAN	GAS CENTRIFUGE	200,000 SWU/YR PROTOTYPE PLANT UNDER CONSTRUCTION
U.S.A. (AVLIS)	ATOMIC VAPOR LASER ISOTOPE SEPARATION	ENGINEERING DEMONSTRATION PLANT UNDER CONSTRUCTION
BRAZIL	BECKER SEPARATION NOZZLE PROCESS	300,000 SWU/YR DEMONSTRATION PLANT UNDER CONSTRUCTION
SOUTH AFRICA	UCOR PROCESS	300,000 SWU/YR PILOT PLANT UNDER CONSTRUCTION
JAPAN	ASAHI CHEMICAL ION EXCHANGE	1-2000 SWU/YR PILOT PLANT UNDER CONSTRUCTION
FRANCE	CHEMICAL EXCHANGE BETWEEN TWO LIQUID PHASES	CHEMEX PILOT PLANT IN OPERATION

Figure 17. Future uranium enrichment plants.

3.2 Centrifuge Plants

Use of the gas centrifuge for uranium enrichment has been under intensive development in the United States since the early 1960's. Emphasis has been placed on increasing machine capacity by using higher peripheral speeds, by build i.g longer machines, and by modifying interna flow distribution to increase efficiency. U.S. machines are pushed to their mechanical limit. Machine capacity order of magnitude greater than Urenco's machines' has been claimed.

Construction of the first two buildings of a gas centrifuge plant at Portsmouth is well along, with initial 2.2 million SWU per year capacity scheduled for 1988. Later expansion to over 12 million SWU per year is possible, by adding six more buildings and replacing machines with advanced centrifuges of higher capacity now being developed.

Figure 18 shows the evolution of centrifuge design up to the present. Figure 19 shows production centrifuges now being tested.

Figure 18. United States gas centrifuges.

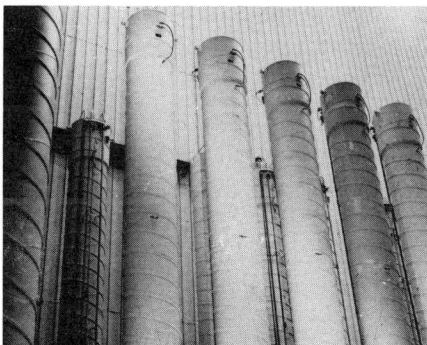

Figure 19. Production centrifuges under test.

An important advantage of the centrifuge is that energy consumption is only 90 kWh/SWU compared with 2200 for diffusion.

Japan also is building a centrifuge prototype plant, with 200,000 SWU per year capacity scheduled for 1985 operation.

3.3 Atomic Vapor Laser Isotope Separation (AVLIS) Process

The Atomic Vapor Laser Isotope Separation (AVLIS) Process, shown schematically in Figure 20, patented by Jersey Nuclear, has been under intensive development by Lawrence Livermore Laboratory and Union Carbide for the U.S. Department of Energy. Uranium metal atoms, vaporized by a beam of high-energy electrons, stream upward between negatively

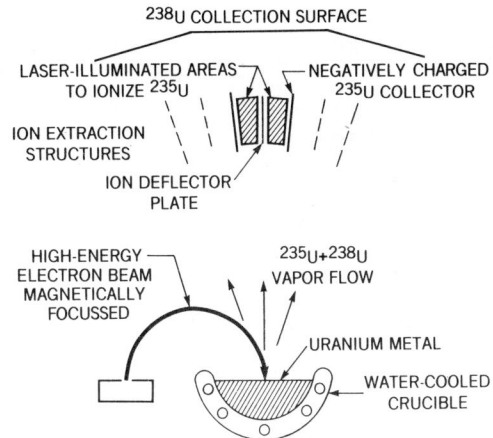

Figure 20. Atomic vapor laser isotope separation (AVLIS) process.

charged plates. The spaces betwen the plates are illuminated by laser light of precise frequencies to excite and then ionize uranium-235 atoms, which are attracted to the charged collector plates. Un-ionized uranium-238 atoms aren't deflected and collect on an upper surface. The AVLIS process module is shown in Figure 21.

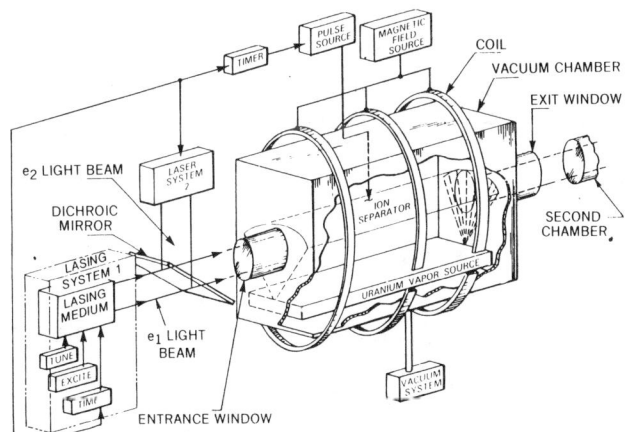

Figure 21. AVLIS process module.

Although the process has difficult problems in handling liquid and gaseous uranium metal and developing lasers of the required frequency, intensity and repetition rate, the Department of Energy has authorized an engineering demonstration of the process in the 1986-1987 period. The favorable cost per SWU projected for the AVLIS process is compared with those for gaseous diffusion and the gas centrifuge in Figure 22.

3.4 Becker Separation Nozzle Process

Dr. E.W. Becker of the Karlsruhe Nuclear Research Center has perfected an aerodynamic process whose principle is shown

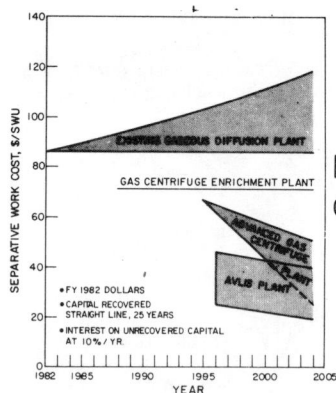

Figure 22. Relative costs of separative work.

Figure 24. Prototype of separation nozzle stage.

in Figure 23. A mixture of 5% UF_6 and 95% hydrogen at atmospheric pressure flows into a low-pressure region through a long, narrow curved slit, or "nozzle." The reduced cross section increases gas velocity, and the curved wall produces a centrifugal field which causes the most sharply deflected gas stream to be enriched in U-235 and hydrogen. A large demonstration cascade of about 300,000 kg SWU per year capacity is being built at Resende, Brazil. Figure 24 shows a prototype of one stage of this plant which has been tested at Karlsruhe. Energy consumption is about 3000 kWh/SWU.

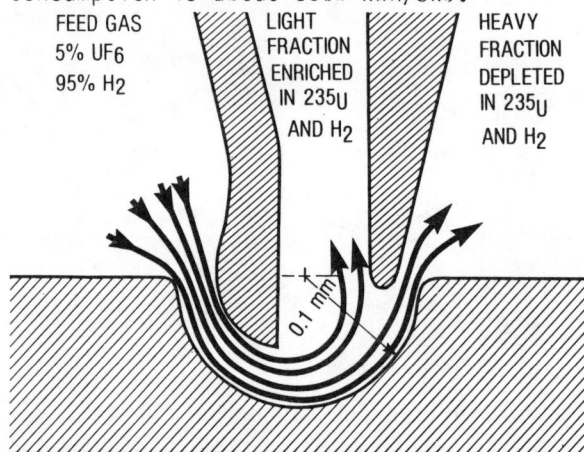

Figure 23. Cross section of separating element in Becker separation nozzle process.

3.5 UCOR Process

A different aerodynamic process, called UCOR, with about the same specific energy consumption as the Becker process, has been developed by the Uranium Enrichment Corporation of South Africa and is used in a large pilot plant being built at Valindaba.

3.6 Other Process

Other processes stated to be competitive with gaseous diffusion and the

gas centrifuge are the Asahi Chemical Ion Exchange Process, for which a one to two thousand SWU per year pilot plant is being built in Japan, and the Chemex process, involving chemical exchange between two uranium valence states in two liquid phases, which has been developed by the French CEA.

4. CONCLUSIONS

This review shows that the monopoly on uranium enrichment once enjoyed by the U.S. gaseous diffusion plants is over. The large Eurodif and Russian diffusion plants and the Urenco gas centrifuge plants now offer separative work at prices comparable to the U.S. However, if the advanced gas centrifuges being developed in the United States live up to their promise and if the Portsmouth centrifuge plant is completed and fitted with these centrifuges, the U.S. will again become the world's lowest cost producer of enriched uranium. And if the AVLIS process lives up to its developer's expectations, the cost of separative work could be reduced still further.

Today's uranium enrichment technology, though very different from the 1940's, is still a challenging and exciting field.

ACKNOWLEDGEMENT

The author acknowledges with thanks the assistance provided by W.J. Wilcox, Jr., and his colleagues at Union Carbide Corporation who supplied many of the figures used in this paper.

NUCLEAR ENERGY—A SECOND VIEW FROM THE OUTSIDE

Edward A. Mason ■ Standard Oil Company (Indiana), Naperville, IL 60540

It's been five years since I was honored to receive the Robert E. Wilson Award in the Fall of 1978. At that time I had been working outside the nuclear industry for about a year and a half. Thus, the invitation to speak here gives me an opportunity to reflect on the use of, and issues connected with, nuclear energy from the vantage point of an additional five years outside the nuclear industry, although I am still very much involved in the energy industry. Those of you who know me personally are aware that I spent the largest part of my professional career at MIT as a professor, researcher and consultant in the fields of nuclear chemical engineering and nuclear reactor safety and engineering. I also served as a member of the Atomic Energy Commission's Advisory Committee on Reactor Safeguards and then as one of the first five Commissioners on the Nuclear Regulatory Commission.

While my present employer is not active in the nuclear field, I have maintained an interest and have had enough personal involvement to follow the progress of the industry, first as a special consultant following the Three Mile Island incident and then for the past three years as a member of the Board of Directors of the Commonwealth Edison Company, which has the largest commitment to nuclear energy of any of our domestic utilities.

Ten years ago in October 1973, OPEC raised the price of petroleum by 70% and the same day nine Arab oil producers cut their production and embargoed the United States and the Netherlands. These two events were but the beginning of a series of abrupt oil supply and price disruptions which have impacted the economies of the entire world. In a number of countries of the world, such as France and Japan, the effect of higher energy prices and uncertainties regarding security of oil supply and predictability of oil price resulted in increased emphasis on nuclear power programs.

All over the world the higher prices for energy have led to energy conservation. This coupled with the current world wide economic recession has caused a decrease in both petroleum and total energy consumption.

In 1978, when I received the Robert E. Wilson Award, the U.S. consumed 78 quads of energy. We were midway through the Carter administration, and were faced with an "energy crisis" which was said to have the "moral equivalent of war". By 1982, U.S. consumption had declined about 9% to 71 quads. This is a vivid demonstration of the impact of both a severe recession, especially in the heavy energy consuming industries, as well as the effect of the conservation efforts undertaken since the mid 1970's. In 1978 petroleum made up almost one-half of U.S. total energy consumption, and of this, half was imported. This meant that about one quarter of our total national energy requirements were supplied by imported petroleum at a cost, at that time, of approximately \$14/bbl. Today petroleum imports have dropped from a high of 8.8 million barrels per day, in 1977, to a low of 2.8 million barrels per day earlier in 1983.

However, the present cost of petroleum at about \$29/bbl is twice that of 1978, so that the energy market still has a tremendous economic impact. In 1982, the cost of the total U.S. energy consumption, on an oil equivalent basis at \$29 per barrel of crude oil, equalled \$1500 for every man, woman and child in our population.

While projections of the rate of growth of energy consumption in our country, in fact in the world, are subject to all the uncertainties which are inherent in making any predictions about the future, there is a long standing historical correlation between energy consumption and per capita Gross National Product. Tremendous progress in energy conservation by individuals and industry as well as the impact of the recession on the level of industrial activity have cut energy growth in recent times. However, we should not fail to take note of the fact that the rate of growth of productivity in the USA has also dropped from 3.3% per year before 1973 to 0.2% in the period 1978-81. If we are to be able to compete effectively in the growing trend toward world markets, this productivity trend must be reversed. American productivity and industrial energy consumption per worker have always been high, and those two related factors have been major contributors to our high standard of living. As we come out of our economic recession, energy consumption will in all likelihood resume its upward course.

It is generally accepted that electric demand will continue to grow faster than the overall rate of energy consumption and faster than the rate of Gross National Product growth, thus increasing the electric industry's share of the energy mix. The experience with operating nuclear power plants has shown that nuclear power is continuing to prove more economical for base load generation than either coal or oil. Furthermore, although there are advanced technologies for electricity generation from "renewable" resources under development which show promise for eventual attractive application in certain areas of the country, these "renewable" energy technologies are still in their infancies. Simply put, there are no proven alternatives to coal and uranium as basic fuels for central station electricity generation over the next few decades.

By the year 2000, about one eighth of electric generating capacity in the United States will need to be replaced. This, coupled with expectations of resumed growth, will require an eventual resumption of the construction of electric generating stations in many areas of the country. Since lead times for construction of generating stations are of the order of ten years, as a nation we must look ahead.

We here today are all interested, in one way or another, in the role nuclear power can play in supplying the electricity needs of the future. Nuclear power plants are used for electricity generation, which in 1982 required about 34% of the primary fuels consumed in the U.S. Electricity generation grew at an average of about 6% per year in the late 1960's and early 1970's. In fact, it was that high growth rate that fueled the demand of nuclear energy and spawned the dreams of cheap, plentiful electricity that many of us recall. However, in the face of a decline of 9% in total energy usage since 1978, electricity output has risen less than 2% and in 1982 declined. Projections of future electricity growth are now in the range of 1 to 2% per year for this country.

Nevertheless, the output of nuclear electricity has grown. We now have 82 nuclear generating units which are licensed to operate in the United States with an aggregate capacity approaching 67,000 megawatts. In 1982, these units produced approximately 280 billion kilowatt hours, or about 12½% of U.S. electric power production. In some areas of the country, the nuclear percentage was much higher. For instance, in 1982 for the Commonwealth Edison Company, which supplies the northern Illinois area, nuclear power accounted for 44% of electric power generation. With the start-up of LaSalle Unit 1 in the past year, Commonwealth's nuclear generation for some weeks this year has reached 55% of the total. In 1982, about twice as much energy from fission was used in the U.S. to produce electricity as from oil and about the same as from natural gas. With a production gain of 47% over the past six years, nuclear power is the fastest growing single segment of U.S. central station electric power production. Indeed, by 1990, when the nuclear capacity now under construction is complete, nuclear power is forecast to account for about 25% of the kilowatt hours generated in the United States.

On the other hand, I'm not telling anyone in this audience anything new when I say that the nuclear power industry is going through a very troubled period. There has not been a nuclear reactor ordered since 1978. Many of those which had been ordered have been cancelled, and a number of those under construction have been dropped. Delays in construction schedules are common and cost overruns are an additional burden to the already financially troubled utilities industry, especially during these days of high interest rates. Public acceptance of nuclear power was dealt a severe blow by the Three Mile Island accident, as well as by the news of the WPPSS construction mismanagement and cancellations. The impact of the WPPSS cancellations on the municipal bond market, as well as problems with quality control during the construction of several nuclear plants are almost daily drawn to the public's attention by the media.

From my vantage point, the most important issue which the nuclear industry must deal with if it is to continue to supply electricity and to grow is that of improvement of the public perception of nuclear power. I believe the technology is far ahead of the publics' perceptions.

The public is "educated" on nuclear energy matters mainly through the media; and the media is in the business of selling themselves wherever possible. Much of the media deal in sensationalism--safety and favorable economics are non-events that simply do not sell newspapers or gain TV audience. In this post-Watergate era of investigative reporting, the nuclear industry must recognize that bad news for them is "good news" for the media. We've all heard such statements in the past. While the industry is now quite aware that they must take an informed, concerted, role in addressing the public's concerns about nuclear matters, for many years this was largely neglected or left to the Atomic Energy Commission.

Many who are in or near the industry repeatedly complain about the uneven or biased reporting of nuclear matters by the media and try to get more opportunities to balance what is perceived as an anti-nuclear bias. Equally important should be efforts by the industry to assure that facts regarding nuclear performance are favorable and that incidents which reinforce any of the public's negative concerns about nuclear power are avoided. The industry needs to do all possible to make the story they wish to bring to the American people factually convincing.

There are three major issues which concern the public, namely: nuclear reactor safety, nuclear waste management, and nuclear power economics. The first two topics have received much attention and I don't have time in this paper to address all three issues. Suffice it to say that the "defense-in-depth" design philosophy in regulatory procedures and actions have been effective in protecting the health and safety of the public even in the face of significant nuclear equipment damage, such as occurred at Three Mile Island. It is still true that, in the United States, we have not suffered a single fatality from radiation through the operation of nuclear power plants--either workers or members of the public. Nor has radiation exposure to the population from normal or even upset conditions, such as occurred at Three Mile Island, increased the annual integrated dose to the population by as much as one per cent over that which occurs from naturally occurring radiation sources (naturally occurring background radiation itself varies by more than 20% from place to place). Furthermore, operator training and qualification is receiving widespread emphasis. Thus, from a factual basis--rather than the fanciful "what-ifs" basis as presented through much of the media by the opponents of nuclear power--nuclear safety, in fact, is far superior to that of other industrial activities. These points can and should be made to the public.

Furthermore, the Congressional passage of the Nuclear Waste Legislation will, hopefully, provide a positive answer to many of the nagging public questions about radioactive waste disposal. This legislation should eliminate an issue which has been exploited by special interest groups to slow nuclear development and add to its costs.

I'd like here to concentrate on the third topic, that of nuclear economics. We are often "informed" through various media that nuclear power is more expensive than alternative forms of electricity production. Frequently references are made to the future, and comparisons are presented of costs of generation from all types of generating technologies and plants which have not even been ordered nor construction started. However, the overwhelming statistics on operating nuclear plants are that nuclear generators are cheaper than coal or oil-fired generators. A survey conducted by the Atomic Industry Forum of 1982 nuclear generating costs recorded by 43 utilities found nuclear to have a 11% cost advantage at the bus bar over coal, and a 56% advantage over oil. This represents the industry average. The experience of

Commonwealth Edison has been even more favorable and serves as an example of what can be accomplished under an effective, concerned management.

Table 1 gives information concerning performance of the nuclear stations on the Commonwealth Edison system during 1981 and 1982. Although Edison's total and nuclear electricity generation declined somewhat during 1982, relative to 1981, due to the curtailment of operations of a number of Commonwealth's largest industrial consumers, the nuclear generation in both years amounted to just under one-half of total generation. The cost savings due to nuclear generation over that which would have occurred if the same energy had been generated from low sulfur coal amounted to an average of about $600 million in both years, and over $2 billion compared with oil. To put a "bottom line" aspect to these numbers, the 10 year savings on the Commonwealth system due to nuclear generation as compared to the use of low-sulfur coal (with which Commonwealth has extensive experience) are $3.2 billion. This is equivalent to almost three times the current investment Edison has in their operating plants which, except for LaSalle I, were installed in the early 1970's.

In 1982 the availability of nuclear and coal plants was essentially the same, but due to the increase in base load usage of nuclear plants the capacity factor for nuclear was higher than that for coal in both years. During the very hot summer of 1982, in which new record peak demands were experienced, the average availability of Commonwealth's six large nuclear units was 92%.

A more detailed breakdown or comparison of the performance of the large nuclear and coal-fired units on the Commonwealth system is given in Table 2. The six nuclear plants[1] were all placed in service between August, 1970 and September, 1974; the six large coal-fired

---

[1]Dresden-2,3; Quad Cities-1,2; Zion-1,2

units were placed in service from 1965 to 1975. Although the 1977 investments in the coal-fired units were approximately 14% less than the average of the six large nuclear units, by 1982 the cost of the equipment added to the coal fired units to provide required environmental controls had increased their average capital cost by 68% whereas backfitting of the nuclear units had increased their average cost by only 20%. Thus, by 1982, the capital costs for the coal-fired units were 21% higher than that of the nuclear units. These are actual plant construction costs and, therefore, reflect facts--not projections. This shows that the capital cost of large nuclear units constructed by a competent utility are very competitive with those of large environmentally acceptable coal-fired plants built at the same period of time.

We often hear it said that although nuclear fuel costs are less than those of fossil fuel, the total power costs are greater. The data in Table 2, based on actual operating experience, show that, in fact, total nuclear generating costs are significantly cheaper than are the coal-fired generating costs. Because of their lower fuel costs, the nuclear plants were operated at higher capacity factors than the coal-fired plants, which favored the nuclear costs of generation. Even eliminating this favorable aspect of nuclear power by evaluating costs at a uniform 60% capacity factor does not change the picture. In 1977, electricity from the nuclear plants was 29% cheaper than from the coal-fired plants and 48% cheaper in 1982--that is, in 1982 coal produced electricity was actually almost twice as expensive as nuclear electricity. This suggests that nuclear power is more inflation proof than coal or oil-fired generation.

Thus we see that, when large fossil units are compared with large nuclear units over the same time frame, nuclear generation enjoys a significant economic advantage, and this can be expected to grow over time.

Table 3 shows how the construction

costs of nuclear units have escalated in the last ten years. A comparison of two comparably sized pressurized water reactor stations, namely, Zion and Byron, which were built approximately ten years apart, is presented. The capital cost for the two Zion units was $280/KWe; ten years later the cost of the Byron station is expected to be $1493/KWe. In addition to inflation in costs of materials and labor, the man hours per kilowatt that went into the Byron station construction were 2.6 times that for Zion; this is an indication of the construction rework required to comply with continually escalating regulatory requirements. The amount of concrete employed in the Byron station is almost double that of Zion and the construction period was 1.7 times that for Zion. This cost escalation is not unique to nuclear construction, since all construction costs were subject to heavy inflation during this ten year period.

A very significant capital cost item, which is a result of the increased time of construction as well as the higher costs of money during the period of construction of Byron, is that of the allowance for funds used during construction. AFUDC was 45% of direct costs in the case of Byron, and only 20% in the case of Zion. This illustrates the economic penalties which result from the stretch-out of construction times in a period of high interest rates.

The construction of nuclear stations and the escalation of the associated costs has been an issue in the public mind for some time, but the experience of WPPSS and Zimmer have intensified attacks on the entire industry based on this issue. I'd like now to address the issue: can nuclear construction costs be controlled? I'll present information which says they can, but aren't always, and suggest that in the public's interest, the nuclear industry, acting collectively, needs to bring its good experience to bear to assure good cost performance everywhere.

The construction and operation of large scale nuclear generating stations has developed only over the past 20 years. Furthermore, this experience has been spread over approximately 60 different utilities; many utilities are building only their first or second nuclear unit. In view of the large capital and operating costs involved as well as the stringent requirements for safety, it is important that experience be shared, not only rationally, but also internationally. In view of the public concerns about the potential hazards from nuclear power, it is important that all reactors be constructed and operated safely since an accident at any one facility arouses fear around the world. The utility industry, both in the United States and internationally, has a number of cooperative programs for sharing information gained from experience. For example, The Steam Generator Owners Group includes utilities from Japan, Sweden, France, Great Britain, Belgium, Italy, and Spain in addition to the United States. This group has been extremely effective in aiding its members in understanding the operation of PWR steam generators and in developing solutions to existing and future problems. The Electric Power Research Institute in the United States has also been very effective in conducting utility-sponsored research and analyses, and in providing advice to electric utilities in the United States. EPRI devotes a considerable portion of its budget to the nuclear field.

Following Three Mile Island, the Institute for Nuclear Power Operations (INPO) was established to aid in nuclear reactor operator training and in providing an industry-wide approach to the safe operation of nuclear reactors. These industry actions are to be applauded since they work not only to the benefit of the individual utilities and the public in their service areas, but also to the industry as a whole by increasing the total amount of information available at any given site.

Another area where increased cooperation could, I believe, be a benefit is that of increased sharing of experience gained in the construction of nuclear

reactors. We hear of wide variations in the effectiveness of nuclear construction management and costs, among the most sensational of which are those of WPPSS and Zimmer. Because they are so extreme, reports of that type invariably find their way into the media and are used unfairly to paint a poor picture of performance by the entire nuclear energy industry.

In an attempt to sort out whether there is any correlation between current costs of construction and construction experience, Figure I was prepared. It presents the estimated nuclear construction costs, in $/KWe, for reactors recently completed or still under construction versus the projected service dates for the units (in the case of multiple units the average service date is used as the abscissa). The data points are differentiated in two different ways by the level of experience of the utility involved. The various data points distinguish between those utilities which now have one or fewer operating units and those utilities which now have two or more operating units. In my mind this represents a rough division in characterizing the degree of nuclear experience that various utilities have. Distinction is also made for plants which are being built at a single unit site versus multiple-unit sites; it is to be expected that there might be some cost advantages and experience gained from building more than one unit at a single site.

One striking thing about Figure I is the wide dispersion of nuclear construction costs for units with the same projected service dates--differences approach a factor of four. Another aspect of the data is that there is a definite upward trend in the estimated nuclear construction costs with projected service date; presumably this reflects anticipated inflation as well as increased regulatory requirements. It is possible to separate the data into three different correlations. The lowest cost line (A) is a least squares fit of the open and solid square data points, which represent reactors under construction at multiple-unit sites by utilities which already have two or more operating units. (Three abnormally high square points are shown crossed out and were not included in the correlation since they are statistically different from the others; these units represent unusual cases as evidenced by the fact that all three units are already seven or more years behind their originally scheduled completion.) The next higher correlation of construction costs (B) represents those utilities which have one or fewer operating units, but are constructing units at multi-unit sites. Finally, the highest cost group (C) represents those nuclear power plants under construction at single-unit sites by utilities which have one or fewer operating units.

Thus, the data in Figure 1 show the construction cost advantages being realized by both experienced utilities and by use of multi-unit sites. This parallels the experience in the construction of other large projects where experience and use of common facilities results in lower costs. Estimates of large coal-fired plants equipped with the type of environmental controls which are required today are comparable with those shown here for the better of the nuclear units.

The larger utilities, namely those which are large enough to have two or more operating units and still have units under construction, have assembled and trained large engineering and construction teams which can better control costs than the smaller utilities who are attempting to build their first nuclear unit on a single site. This has certainly been the case for Commonwealth Edison; their units still under construction are identified in Figure I by the solid squares. This is not due to luck, but to organization, management, and experience. Table 4 shows the significant size and level of education of their nuclear staff. Not only does Commonwealth's scale of operation enable them to have a large and highly educated staff, but they recognize the rigor required for effective nuclear generation and have the management commitment to follow through. Commonwealth Edison is acting as its own general contractor on all units under construction so as to have firm control; their Manager of Nuclear (construction) Projects reports directly to the Chairman; all nuclear operations and staff report into the executive offices of the company, independent of non-nuclear operations. Nuclear quality assurance reports independently to the Vice Chairman.

Edison's record suggests that good management and experience can and has produced superior results, as it does elsewhere. Regional pooling of construction of nuclear power plants and/or transfer of experience from the experienced utilities to the smaller, or less experienced utilities, should therefore be of great benefit in controlling costs and quality of construction, and in so doing would improve the entire industry's public image. Application of these kinds of management techniques and construction experience should be helpful in countering statements which often appear in the media that, across-the-board, nuclear construction costs are too high and are uncontrollable.

The excellent proven safety record of nuclear power plants--the significant economic advantages over both coal and oil--the almost complete absence of environmental impacts (such as acid rain) due to the closed cycle nature of the plants--all are factors that argue towards the continued, and where justified by load growth, the increased usage of nuclear power. At the present time the utilities are faced with difficult financing problems, and projections of load growth are so uncertain that utilities are very reluctant to undertake the construction of any new large electric generating stations. These factors, coupled with the regulatory and scheduling uncertainties which arise from publically expressed concerns regarding nuclear power, have brought many utilities to the point of not even including consideration of nuclear power in their planning. Yet other countries, such as France and Japan, proceed. In view of the real and already demonstrated long term advantages of nuclear power to this country, it is vital that the industry work together in many and imaginative ways to assure that nuclear continues to be, in fact, the safest, cheapest, and environmentally most benign method of electricity production available. The industry needs to work to assure that good management practices and knowledge based on experience find widespread adoption in the industry. Such efforts are necessary to persuade the rational public that the facts regarding nuclear energy, not fear based on fancy, should be used in public decisions regarding energy choices.

EAN/cl

TABLE 1

NUCLEAR POWER HIGHLIGHTS
COMMONWEALTH EDISON COMPANY

	1981	1982
GENERATION - BILLIONS OF KWH	27,089	25,512
% OF TOTAL GENERATION	45	44
SAVINGS OVER LOW-SULPHUR COAL $/MILLION	540	615
SAVINGS OVER OIL $/MILLION	2,000	2,400
% AVAILABILITY - NUCLEAR	77	72
- COAL	66	75
% CAPACITY FACTOR - NUCLEAR	64	59
- COAL	45	51

TABLE 3

NUCLEAR CONSTRUCTION COST COMPARISON

	ZION	BYRON
MANUFACTURER	WESTINGHOUSE	WESTINGHOUSE
CAPABILITY, MWE	2,080	2,240
SERVICE DATES	10/73, 9/74	6/84, 11/85
CAPITAL COSTS-($/MILLION)		
DIRECT	473.6	2,224.7
INDIRECT	107.9	1,120.0
TOTAL	581.5	3,344.7
$/KWE	280	1,493
AFUDC-% DIRECT COST	19.8	45.1
MANHOURS PER KW		
CONSTRUCTION	5.3	13.7
ENGINEERING	N/A	3.8
YARDS OF CONCRETE	270,000	485,000*
MILES OF PIPE	100	139
MILES OF CABLE	1,200	1,700
CONSTRUCTION PERIOD (MONTHS)	67	114

*EXCLUDES COOLING TOWERS

TABLE 2

COMPARISON OF PERFORMANCE
OF
LARGE COAL FIRED AND NUCLEAR UNITS
COMMONWEALTH EDISON COMPANY

	6 LARGE COAL-FIRED UNITS		6 LARGE NUCLEAR UNITS	
	1977	1982	1977	1982
CAPITAL COST-$/KW	174	292	202	242
NET CAPACITY FACTOR-%(C.F.)	46.2	51.4	61.0	58.9
ENERGY COSTS-(MILLS/KWH)				
CAPITAL CHARGES	8.4	13.0	7.5	9.4
OPERATION & MAINTENANCE	2.4	4.6	2.1	5.1
FUEL	10.1	27.9	3.5	8.3
TOTAL	20.9	45.5	13.1	22.8
TOTAL @ 60% C.F.	18.7	43.3	13.2	22.4
NUCLEAR ADVANTAGE-(%)	BASE	BASE	29	48

TABLE 4

NUCLEAR STAFF
COMMONWEALTH EDISON COMPANY

ADVANCED DEGREES	107
DEGREED	651
OTHER*	2357
TOTAL	3115

LICENSED TO OPERATE NUCLEAR REACTORS 339
NUCLEAR AND FOSSIL STAFFS SEPARATE
* INCLUDES TECHNICAL AND NON-TECHNICAL STAFF

Estimated nuclear construction costs

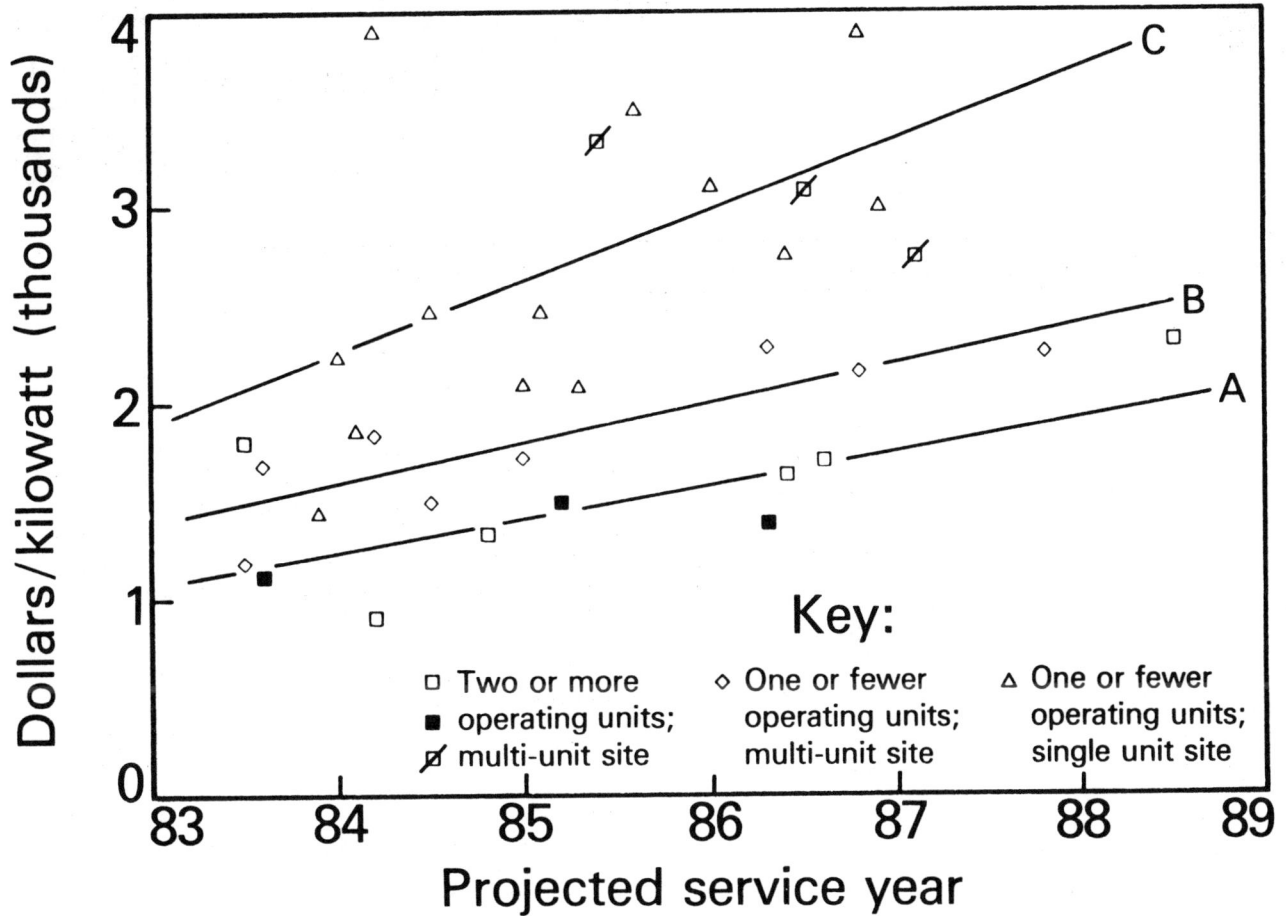

Figure 1.

STATUS AND FUTURE PROSPECTS FOR NUCLEAR FUEL REPROCESSING

James A. Buckham ■ President, Allied-General Nuclear Services, Barnwell, SC

INTRODUCTION

In 1974, in response to receiving the Nuclear Engineering Division's Robert E. Wilson Award, I asked what seemed a rhetorical question: "Why reprocess nuclear fuels?". At that time, there were seven good reasons to proceed with reprocessing. The holdup in doing so in the United States appeared to be government procrastination on establishing criteria for handling plutonium recycle and final disposal of high-level waste.

Shortly thereafter, following the announced explosion of a nuclear device in India, there became increasing concern that reprocessing might lead to the proliferation of nuclear weapons. This issue even became the subject of comments by presidential candidates in the 1976 election.

Today, the same seven reasons to proceed with reprocessing are as valid as ever, and in fact, an eighth reason has emerged. However, in the early months of the Carter Administration, following

further procrastination on establishing the needed criteria, nuclear reprocessing was hit with a severe sledge hammer-like blow in the form of a Presidential ban on commercial reprocessing. This ban was intended to provide time to study the proliferation issue and to set an example to the rest of the world that foregoing reprocessing could have worldwide benefits. After a two-year, sixty-nation study, it was clear that the United States was virtually alone in its position. The rest of the nuclear nations made it clear they intend to proceed with reprocessing despite the U. S. position. In fact, to many, the U. S. position appeared to be counter productive.

Although the Reagan Administration subsequently lifted the ban on reprocessing, thus removing the sledge hammer, it has as yet done nothing to heal the wounds that this blow created. This leaves nuclear fuel reprocessing in the United States precariously close to extinction, and largely for politically inspired --- rather than technical or economic --- reasons. Reprocessing is the

Allied-General Nuclear Services, Barnwell, South Carolina.

corner-stone of extending nuclear fission resources beyond the short period of a few decades, for with reprocessing, mankind would have access to nuclear fission energy for many thousands of years. That, together with the inherent energy self-sufficiency reprocessing provides, is the reason other nations are proceeding even though we are not. In this paper, I will first recount the reasons for proceeding with reprocessing, and then discuss the technical capability for doing so, the economic factors associated with it, the political issues that remain, and finally, offer a prognosis of the future.

REASONS FOR REPROCESSING

Two of the reasons for reprocessing relate primarily to its superiority in connection with waste management. Five of them relate primarily to energy resource recovery. The eighth relates to a special logistical problem created by the procrastination. I will comment on these reasons in that order.

1. Better Waste Management: The alternative to fuel reprocessing is to dispose of spent nuclear fuel as waste directly after one-time usage in the reactor. Compared to such spent fuel disposal, reprocessing is superior both from environmental and economic points of view. By removing virtually all of the plutonium from the spent fuel, a significant long-term radioactive element is eliminated from the waste. This makes the waste less hazardous; it also permits disposal of plutonium as a fuel, eliminating it from the face of the earth. In addition, waste disposal safety is improved because the remaining fission products can be converted to almost any desired chemically stable form instead of remaining in the chemically-reactive, partially oxidized state in which they were formed during the fission process. Borosilicate glass is a current front-running candidate for the preferred chemical corm. However, many other forms being considered

offer various advantages and disadvantages compared to glass. Any of these forms would be vastly superior to that of spent fuel.

Two factors lead to economic improvements in waste management. First is the significantly reduced volume of glass, which would be approximately one-tenth that of the equivalent canned fuel assembly. Some consideration is being given to physical disassembly and recanning of the fuel rods from fuel assemblies prior to disposal. While this could create a volume reduction as high as three-fold, the resulting volume would still be some three-fold greater than glass. The second cost saving factor, which should be realizable, is that reprocessed waste need not be entombed retrievably whereas both public law and common sense require spent fuel to be so entombed at inherently greater cost.

Admittedly, other solid wastes stem from fuel reprocessing. These include a variety of low heat generating materials such as cladding hulls, ash from incineration of combustible trash, and miscellaneous contaminated equipment and tools. While this volume approaches that of the original spent fuel, its disposal is not costly because it could be used as backfill in the repository tunnels after the high-level waste has been placed in the floors or walls of those tunnels. Thus, it should be possible to dispose of this relatively low-hazard material for only the cost of transportation and handling.

2. Continuing Improvements in Waste Management: Disposal of spent fuel as waste provides little option for improved practice in the future. Only by proceeding with reprocessing are improvements in the safety and cost of waste management likely. One can envision many alternatives evolving in chemical processing flowsheets after adequate experience is obtained with

waste disposal that would lead to smaller volumes, less reactive substances, or other improvements. One such alternative of particular appeal might be to segregate high-level reprocessing wastes into various fractions according to their relative hazard and to then subject each to an appropriately different treatment. For example, remaining transuranics could be segregated on the basis that they are the really long half-life materials. Elements such as cesium and strontium could be segregated as the largest heat producers in the waste. Only through reprocessing could such improvements be effected.

3. Recycle of Plutonium: I have already mentioned that plutonium recycle has an advantage from the standpoint of better waste management. There is also an advantage in using the plutonium as a fuel in that it extends uranium resources considerably.

Contrary to some statements, the hazards for using recycled plutonium in light-water reactors are not significantly different from those in current operations. During the last six months or so that fuel is in a reactor, approximately half of the energy produced comes from fissioning of the plutonium produced in that reactor. In fact, half of the plutonium produced in the reactor is fissioned, and thus destroyed, before the fuel is removed.

The most vocalized deterrent to the recycle of plutonium is the perception that it might be diverted to weapons rather than being used as fuel. This perception is not particularly well-founded. In the first place, fuel-grade plutonium makes a very poor weapon because of its very high radioactivity compared to conventional weapons-grade plutonium. In fact, it appears that no weapon has ever been made from fuel-grade plutonium nor would it be the choice of any

weapons-maker to do so. Nonetheless, safeguards systems are believed required to assure that no such diversion occurs. Such systems have been in development, particularly in the U. S., during the past decade. A highly advanced safeguards system for a reprocessing plant was developed at the Barnwell Nuclear Fuel Plant under contract to DOE during the past five years. This system basically provides physical security methods that prevent unauthorized performance of any action that could possibly be associated with diversion. In addition, the system provides for computerized hourly inventories of plutonium to replace the former annual clean-out and analysis-type inventories. Thus, IAEA inspectors at any plant with such a system would receive accurate and timely warnings of a diversion possibility and could initiate prompt and appropriate response.

4. Recovery of Unused Uranium: Spent fuel contains about 96% uranium, about 1% of which is the fissionable isotope U-235. This uranium is directly reusable by mixing with current feedstocks to enrichment plants and is substantially more valuable than natural uranium because of its higher U-235 content.

5. Permits Effective Use of Breeder Reactors: Breeder reactors are under development in the U.S. and overseas; in fact, large-scale breeders are already in service in France. The purpose of a breeder reactor is to produce (from U-238 as plutonium) more fuel than is consumed. Eventually, most of the U-238 on earth would be available for fuel; this would multiply by sixty or seventy fold the amount of energy we get from our uranium resources. With breeders, reprocessing is necessary because it serves no purpose whatsoever to breed fissionable material unless it is recovered and used. For satisfactory performance, initial breeder cores should be fueled with high-burnup

plutonium that can only be recovered from commercial spent fuel. Thus, reprocessing should be a precursor of breeders.

6. Permits Use of Thorium Ore: Thorium, which is much more abundant in nature than uranium, can be converted to fissionable U-233 in a nuclear power reactor. One type of reactor using such fuel is the high temperature gas-cooled reactor, and other systems have been considered. Reprocessing, of course, is again vital for successful use of thorium ore in that it is a necessary step in recovering the bred U-233.

7. Fission Product Utilization: Beneficial usage of fission products has long been a dream of many. I frankly cannot recommend wide-spread use of radioactive fission products because this is contrary to sound waste management practice. However, some fission products are not radioactive and are very rare in nature. It is from these that I think the initial use of fission products might come. I would like to cite, as an example, recovery of the platinum family of metals (rhodium, palladium, and ruthenium) that are present in spent fuel as largely non-radioactive isotopes. Platinum metals not only have an increasing use as catalysts in industry, as well as in exhaust systems of automobiles, but also are in relatively short worldwide supply. The United States now imports essentially its entire requirements from South Africa and the Soviet Union. Recovery of these and other fission products from spent fuel requires proceeding with reprocessing and could be adopted, at least in new plants, as future needs become more clear.

8. Precludes Costly Interim Storage: Because most of the nuclear reactors which are currently in operation were expected, when they were designed, to have promptly available reprocessing services, they do not generally have adequate storage for spent nuclear fuel. For these reactors, some form of additional fuel storage will be needed if reprocessing does not proceed promptly. Because of the lack of certainty about reprocessing, newer reactors have been designed with fuel storage pools sufficient for the lifetime operation of the reactor. Nonetheless, an overall problem of insufficient fuel storage faces the nuclear electric utilities. The Nuclear Waste Policy Act of 1982 calls for initial operation of a Federal terminal waste disposal repository in 1998. Even if that schedule is met, the repository would be receiving spent fuel (if that were the disposal waste form) at a rate less than it is being produced in the country. It is thus readily apparent that the initial repository will not solve the nation's spent fuel storage problem. Conversely, any fuel that is reprocessed prior to 1998, and to some considerable extent beyond that date, will obviate the need for building costly additional interim fuel storage facilities, either at the power plants or in some centralized location. Any money spent on such interim storage, in the long run, is wasted money and represents a penalty we are forced to pay for past indecision.

TECHNICAL FACTORS

With eight very sound reasons for proceeding with reprocessing, and with no significant forward movement in the United States, one might surmise that we don't know how to do it. Nothing could be further from the truth. Reprocessing plants have been operated for the benefit of the government in connection with weapons material production in the United States for over 40 years. All reprocessing plants currently in operation throughout the world use the Purex process. Purex, now used for about 30 years, is a well-proven process which employs fuel dissolution followed by solvent extraction to separate dissolved

fuel into its major constituents ---
uranium, plutonium, and fission products
or high-level waste. All plants today are
using or converting to the use of the same
solvent, tributyl phosphate dissolved in a
high-quality kerosene, and the years of
successful experience with the solvent are
very reassuring.

Commercial power reactor fuel has been
reprocessed in an initial small pioneering
plant in the United States, in pilot
plants in many foreign nations, and in
larger scale facilities in France and the
United Kingdom. Currently, France and the
United Kingdom are constructing large
modern reprocessing plants for commercial
fuel and design efforts are underway in
Germany and Japan to do the same.

At Barnwell, South Carolina, the
United States has one plant nearly
completed which would be capable of
reprocessing commercial spent fuel. Two
earlier efforts at reprocessing in the
United States were undertaken; one at West
Valley, New York and one in Morris,
Illinois. The West Valley plant, the
pioneering plant I previously mentioned,
reprocessed several hundred tons of spent
commercial nuclear fuel before being shut
down for modifications and expansion.
However, during that shutdown, it became
apparent that new seismic and tornado
criteria would make the cost of
modernizing the plant greater than the
cost of starting anew; consequently, that
plant has since been permanently closed.
The small plant in Morris was designed to
utilize an innovative process and
proceeded through cold-testing; however,
here too it was found that the cost of
required modifications would exceed the
cost of starting anew. Hence, the Morris
Plant has never operated.

The Barnwell Nuclear Fuel Plant
(BNFP), which could reprocess commercial
fuel in this country, is a large modern
plant, having a capacity of 1,500 metric
tons per year of spent fuel --- sufficient
to serve 50 to 70 large nuclear power
plants. Construction of the BNFP's basic
Separations Facility and an associated

facility to convert recovered uranium to
the hexafluoride had been largely complete
in 1976 when licensing hearings were
postponed pending completion of generic
hearings on the environmental consequences
of recycling plutonium. Then, in 1977,
after President Carter banned commercial
reprocessing, the NRC terminated all
plutonium-related licensing proceedings.
No subsequent regulatory consideration has
been given to the recycle of plutonium. If
the political factors (and economic
factors arising therefrom) could be
resolved, the Barnwell plant could
successfully reprocess spent fuel for most
of the existing U.S. reactors and for
approximately one-third of those expected
to be in operation in the U.S. at the end
of the century. The BNFP, in fact,
represents the only chance for commercial
fuel reprocessing to proceed in the United
States in this century and probably early
into the next century.

ECONOMIC FACTORS

Full-scale reprocessing of spent
nuclear fuel is estimated to cost between
$300 and $500 per kilogram of uranium
(KgU) contained in the spent fuel. These
costs would need to be offset in a
commercial enterprise by cost savings or
revenues from the following sources:

1. Savings of costs otherwise arising
    from interim spent fuel storage.

2. Savings in the lower costs of disposal
    of waste versus those for disposal of
    spent fuel.

3. Revenues from sale of uranium
    hexafluoride for recycle.

4. Revenues from sale of plutonium
    product for fueling breeders or for
    recycle in existing reactors.

If proper savings and revenues from
all four sources could be applied,
commercial reprocessing would be a
modestly attractive venture today even
though uranium (and hence plutonium)
values are considerably lower than once

envisioned. Because far fewer nuclear reactors have been constructed than once anticipated, an over-production of uranium exists and a consequent reduction in price has occurred. However, inevitably, when the current over-production is used up and when more dilute ores have to be attacked, the value of uranium and plutonium will rise significantly. At that time, reprocessing would surely be economically very attractive.

Unfortunately today, savings and revenues can be counted on from only two of the four sources --- avoidance of costs for interim storage and revenues from uranium sales. Under these conditions, full-scale reprocessing is not economic. What has been proposed by many, including the present owners of the Barnwell Plant, is a period of demonstrational operation during which the BNFP is donated or otherwise obtained at little cost and operated at a reduced capacity. Under those conditions, even if operated at as little as one-third of capacity, the revenues and savings from these two sources more than offset the estimated reprocessing cost of slightly under $200 per KgU.

Availability of fuel for reprocessing plants is not a problem, at least near-term. As indicated earlier, the BNFP could not keep up with the output of spent fuel even from current nuclear power plants and by the year 2000, even if no new orders for power plants occur, it would take about three Barnwell-size plants to keep up with the load and slowly work off the backlog.

The impediments to proceeding are thus neither technical nor economic, but, instead are associated with political uncertainties.

POLITICAL FACTORS

Arguments against reprocessing in the United States, based upon concern over non-proliferation, have lost their effectiveness with the realization that reprocessing is going to happen anyway in the rest of the world and that therefore our involvement cannot possibly lead to greater proliferation. Indeed, many believe, as I do, that U.S. involvement in reprocessing is very necessary in order to assure that reprocessing conducted overseas is performed with adequate safeguards. By far the biggest remaining political impediment is the government's insistence that reprocessing be conducted totally by the private sector. In turn, the private sector is understandably wary of investing in a business which can be turned on or off by the government for political reasons. Thus, one absolute requirement for industry investment would be some kind of indemnification for full-cost recovery in the event of another future reversal of government policy. At least to date, it has not been easy to find a means to provide such indemnification.

Another serious remaining obstacle is the continuing total lack of resolution of the plutonium usage issue. At the moment, the generic hearings on plutonium recycle remain closed. The NRC has stated that it would take five years and a lot of money to reopen those hearings and bring them to an orderly conclusion. Moreover, the NRC has not yet taken any steps to initiate this. It is absolutely clear that no commitments will be made privately to construct fuel fabrication plants for recycling plutonium as fuel until this plutonium usage issue is fully resolved. Moreover, the development of an impetus to resolve the plutonium issue faces the major obstacle that there is no real urgency for full-scale commercial reprocessing. For the eight reasons which I have discussed, reprocessing is necessary in the long run if nuclear energy is to survive. However, it is possible, though not prudent, to forego reprocessing for at least ten and perhaps twenty years. Thus, there is no readily apparent crisis or pressing reason for industry to take large risks --- either economic or political --- or to undertake heroic measures to implement reprocessing.

However, if the technology of

reprocessing is to be nurtured for long-term realization of its benefits, prompt action would indeed be prudent. This is due to the rapidly diminishing availability of the BNFP, which is in the process of being shut down by its current owners. Lacking other direct interests in nuclear energy, events of the last seven or eight years have destroyed any desire on the part of the plant's present owners to proceed with reprocessing as a business. Yet they have repeatedly stated their flexibility and willingness to make the plant available for demonstrational operation to the nuclear utilities and suppliers who do have a genuine long-term stake in nuclear energy. Unfortunately, with the many, many problems that American utilities are facing, it is understandably difficult for them to come to grips with the reprocessing issue in the absence of highly visible and urgent need.

THE FUTURE?

There is no possibility for commercial reprocessing of spent nuclear fuel in the U. S. for at least twenty years without the BNFP. The current owners are presently in the process of shutting down the BNFP, cleaning out from its equipment the natural uranium used in cold-testing, and selling peripheral equipment to offset the clean-up and shutdown costs. While the reprocessing buildings, with their concrete cells and stainless steel and titanium equipment, will be left in place and should not deteriorate, it will still be very difficult and much more costly to proceed with reprocessing by reactivating a plant that has been unused for many years than to preserve it and initiate limited-scale, demonstrational reprocessing at an early date. A possibility for this exists through an effort being made by several U.S. nuclear industry and utility organizations to undertake the active preservation of the BNFP during the next few years while a solution to the political problems which confront reprocessing can be sought. Unless such a proposal emerges, reprocessing will not happen in the U. S. in this century and the citizens of the

next century will be the ultimate losers, facing at best a less certain energy future.

SYMPOSIUM SERIES

ISBN 0-8169-0323-9